Comparative Philosophy of Religion

Volume 2

This book series publishes works of comparative philosophy of religion—works that are religiously inclusive or diverse, explicitly comparative, and critically evaluative. It serves as the primary publishing output of The Comparison Project, a speaker series in comparative philosophy of religion at Drake University (Des Moines, Iowa). It also publishes the essay collections generated by the American Academy of Religion's seminar on "Global-Critical Philosophy of Religion." The Comparison Project organizes a biennial series of scholar lectures, practitioner dialogues, and philosophical comparisons about core, cross-cultural topics in the philosophy of religion. A variety of scholars of religion are invited to describe and analyse the theologies and rituals of a variety of religious traditions pertinent to the selected topic; philosophers of religion are then asked to raise questions of meaning, truth, and value about this topic in comparative perspective. These specialist descriptions and generalist comparisons are published as focused and cohesive efforts in comparative philosophy of religion. Global-Critical Philosophy of Religion is an American Academy of Religion seminar devoted to researching and writing an undergraduate textbook in philosophy of religion that is religiously inclusive and critically informed. Each year the seminar explores the cross-cultural categories for global-critical philosophy of religion. A religiously diverse array of essays for each seminar are published along with a set of comparative conclusions.

More information about this series at http://www.springer.com/series/13888

Timothy D Knepper • Lucy Bregman
Mary Gottschalk
Editors

Death and Dying

An Exercise in Comparative Philosophy of Religion

Editors
Timothy D Knepper
Department of Philosophy and Religion
Drake University
Des Moines, IA, USA

Lucy Bregman
Religion Department
Temple University
Philadelphia, PA, USA

Mary Gottschalk
Department of Philosophy and Religion
Drake University
Des Moines, IA, USA

ISSN 2522-0020 ISSN 2522-0039 (electronic)
Comparative Philosophy of Religion
ISBN 978-3-030-19299-0 ISBN 978-3-030-19300-3 (eBook)
https://doi.org/10.1007/978-3-030-19300-3

This Springer imprint is published by the registered company Springer Nature Switzerland AG.
The registered company address is: Gewerbestrasse 11, 6330 Cham, Switzerland

Preface

Much has happened with The Comparison Project (TCP) since the publication of its first volume of essays in comparative philosophy of religion: *Ineffability: An Exercise in Comparative Philosophy of Religion* (Springer, 2017). Most notably, TCP published a student-written photo-narrative about religion in Des Moines, Iowa: *A Spectrum of Faith: Religions of the World in America's Heartland* (Drake Community Press, 2017). *Spectrum* was in many ways the product of a digital storytelling initiative about local, lived religion and in turn gave rise to the new initiatives in local, lived religion: a monthly open-house series, an annual interfaith youth leadership camp, and collaborative photo-narrative projects with Minzu University of China and the University of KwaZulu Natal.

This is to say that The Comparison Project is no longer just "an exercise in comparative philosophy of religion" (as we said in the Preface of *Ineffability*). In fact, our work with "local, lived religion"—whether in Des Moines, Beijing, or Pietermaritzburg—often dwarfs our efforts in "global, comparative religion." Nevertheless, our lecture and dialogue series remains the heart and soul of The Comparison Project.

The design of the lecture and dialogue series is still quite simple. Choose a topic that is interesting to our local audiences and important for comparative philosophy of religion, refining that topic through a set of questions. Invite scholars to lecture on the topic from the perspective of different religious traditions, texts, and thinkers. Arrange for local dialogues and special events as well. Then, compare over the content of the series, raising philosophical questions of meaning, truth, and value about the topic in comparative perspective.[1]

In the case of the 2-year series on ineffability (2013–2015), the topic of course was ineffability, with the questions concerning *what* is allegedly ineffable, *how* that ineffability is linguistically expressed, *what reasons* are provided in defense of that ineffability, and *what ends* are at play in that expression and defense of ineffability. In the case of the 2-year series on death and dying (2015–2017), however, we did

[1] For more details about our methods, see the Preface to our first volume, *Ineffability* (Knepper and Kalmanson 2017).

not flag questions as much as stake territory: the relationship between, on the one hand, traditional theologies of death and rituals of dying and, on the other hand, advances in western medicine and the growing medicalization of death. For better or worse, therefore, the specialist essays and comparative conclusions generated by the series on death and dying—which are contained below—reflect more diversity in terms of their argumentative ends: some engage in philosophical bioethics, others in theological apologetics, and still others in anthropological description (more about this in the Introduction).

None of this would have been possible without generous funding. In the case of our 2015–2017 lecture and dialogue series, we thank Drake University's Center for the Humanities; Drake University's Principal Financial Group Center for Global Citizenship; the Medbury Fund; Humanities Iowa; the Des Moines Area Religious Council; Cultivating Compassion: the Dr. Richard Deming Foundation; and Iles Funeral Homes, our special sponsor for the series on death and dying. None of this also would have been possible without the people of The Comparison Project: its directors, Tim Knepper (Professor of Philosophy, Drake University) and Leah Kalmanson (Associate Professor of Philosophy, Drake University); its specialists for the 2015–2017 series, Lucy Bregman (Professor of Religion, Temple University), Allen Zagoren (Associate Professor of Practice in Public Administration, Drake University), and Mary Gottschalk (adjunct professor, Drake University); and its steering committee, students Isaiah Enockson and Anoushe Seiff, and community representatives Sarai Rice (executive director of the Des Moines Area Religious Council, Richard Deming (founder and chairman of Above + Beyond Cancer), Ted Lyddon-Hatten (director of Drake's Wesley House), and Mary Gottschalk (yet again).

Des Moines, IA, USA — Timothy D Knepper
Philadelphia, PA, USA — Lucy Bregman
Des Moines, IA, USA — Mary Gottschalk

Reference

Knepper, Timothy D., and Leah E. Kalmanson, eds. 2017. *Ineffability: An exercise in comparative philosophy of religion*. Cham: Springer.

Contents

Chapter 1
Introduction: Death and Dying in Comparative Philosophical Perspective

Timothy D Knepper

Abstract This introductory chapter previews the content and conclusions of The Comparison Project's 2015–2017 programming cycle on death and dying. First, it explicates each of the 12 content essays on death and dying, especially with respect to the focus of the series—how traditional theologies of death and rituals of dying are affected by and respond to the medicalization of death. Then, it highlights the four, brief comparative conclusions to the series.

1.1 Content

Who or what dies? How and why? What if anything comes after death? For whom or what? When and where? How ought we die? How should we ritualize dying? What should be done with dead bodies? How should we remember the dead?

Humans have asked so many questions about death and dying and done so from so many different perspectives and backgrounds. In the case of traditional philosophy of religion, these questions have usually been limited to the survival of death. This "exercise in comparative philosophy of religion," by contrast, focuses instead on the question "how should we die?"—especially as traditional religious answers to this question have contended with the increasing tendency, at least in "the West," to "medicalize dying." As an exercise in comparative philosophy of religion, this volume attempts not just to describe and compare but also to explain the patterns that emerge in our comparisons and to offer tentative, localized evaluations regarding "how we should die."

This is of course no easy task; it is, though, one that begins simply—with the religious traditions and communities of the world. In the 2015–2016 and 2016–2017 academic years The Comparison Project explored theologies of death and rituals of dying through 12 scholar lectures, four community dialogues, a set of ritual

T. D. Knepper (✉)
Department of Philosophy and Religion, Drake University, Des Moines, IA, USA
e-mail: tim.knepper@drake.edu

T. D. Knepper et al. (eds.), *Death and Dying*, Comparative Philosophy of Religion 2, https://doi.org/10.1007/978-3-030-19300-3_1

performances, an evening of creative non-fiction readings, and an array of concluding comparisons. The lectures and conclusions appear here, as rigorous academic essays, along with two supplemental essays about death and dying in traditions that we did not cover in lecture form during the series.[1] Collectively, 11 different religious traditions are present (depending on how one itemizes religious traditions), as well as secular perspectives on death and dying.[2]

Although the title of the 2015–2017 series was simply "death and dying," we asked lecturers to focus on the challenges of medicalized dying for traditional theologies of death and rituals of dying.[3] In some cases, this yielded detailed religious bioethical systems; in other cases, efforts to actively confront medicalized dying; in yet other cases, simply a consideration of what religious texts, thinkers, or traditions would say about medicalized dying. For the sake of the coherence of this volume, we decided to organize the content essays according to these headings rather than in the chronological order in which their corresponding lectures were delivered. Thus, the first section contains essays about "death and religion"; the second section, "medicalization and religion"; and the third section, "bioethics and religion." The final section of the collection then contains four concluding efforts to philosophize about death and dying in comparative perspective.

The first essay of Part I ("Death and Religion"), Mark Berkson's "Death in Ancient Chinese Thought," tells us "what Confucians and Daoists can teach us about living and dying well," especially where "us" includes those who attempt to find meaning and consolation in the face of finitude and death without recourse to the concepts of an afterlife and personal survival of death. Focusing primarily on the early Confucian Xunzi and early Daoist Zhuangzi, Berkson shows how the former helps us "find meaning within the pain of grief and loss, more deeply understand the power and efficacy of death rituals, and more fully appreciate the nature of our relationships with the dead," while the latter "challenges us with the thought-provoking possibility that we can transform ourselves in such a way so as to avoid grief and accept mortality with equanimity and humor." Both, therefore, not only "offer a critique of an overly medicalized approach to dying" but also "present visions of what constitutes a good death."

[1] For obvious reasons, we did not include here the four interfaith dialogues involving representatives of local religious communities, the performance of Tibetan Buddhist rituals related to death and dying, and the reading of non-fiction essays by cancer survivors. The supplemental essay by M. Allison Prude on "Death in Tibetan Buddhism" replaces the Tibetan Buddhist ritual performance with an academic analysis of theologies of death and rituals of dying in Tibetan Buddhism, while the supplemental essay by Mark Berkson on "Death in Ancient Chinese Thought" provides coverage of a lacuna in our series.

[2] Our secular perspective, as reflected in the essay of Amy Hollywood, is in no way an attempt to develop a secular bioethics. Instead, it is an imaginative exploration of the role that "difficult literature" might play in the preparation for death of those who are no longer religious. For the attempt to develop formal, secular bioethics, see especially Engelhardt (1991).

[3] Our understanding of the medicalization of death refers to the substitution of doctors, hospitals, and technology for family, home, and comfort during the process of dying. More fundamentally, it views death as a purely biological phenomenon rather than a cultural, emotional, and religious experience that is a natural part of human existence.

Amy Hollywood's "Secular Death" next explores how those who live outside traditional religious communities might approach death—both their own and those of others. This exploration enters into conversation first with Susan Howe's book-length poem *That This*, which was dedicated to the memory of her third husband, then with David Orr's *Beautiful & Pointless: A Guide to Modern Poetry*, which was begun as his father was dying of cancer. It is in reading and writing such "difficult literature" that Hollywood finds a "training ground" for how the difficulty of death might be approached by "those who are no longer Christian."

Michelene Pesantubbee's "Negotiating Advance Directives in a Navajo Context" looks at recent initiatives to encourage Advance Directives among the Navajo. Pesantubbee first examines Navajo conceptions of death and practices of dying, especially insofar as they have been at the crux of clashes with hospital treatment. Although misunderstanding between western and traditional medicine began to alleviate in the 1990s—in large part due a recognition of the importance of cultural traditions to Native American health care—this "cultural-sensitivity training" ironically increased hospital reticence about end-of-life discussions with the Navajo. Only in 2006, with Dr. Timothy Domer's home-based care program at Fort Defiance Hospital, did things begin to change due to the creation and use of a Navajo poem about death that served to connect traditional Navajo concepts of death and state health directives. For Pesantubbee, the moral of the story is that cultural-sensitivity training pays off in the long run, albeit in conjunction with accommodation for traditional beliefs and practices.

The final essay of Part I—"The Cult of Santa Muerte: Migration, Marginalization, and Medicalization"—is unique in that it draws together two different lectures on the "cult" of Santa Muerte. Why two different lectures on Santa Muerte? Suffice it to say that institutional exigencies provided the opportunity to host lectures on Santa Muerte each year in conjunction with Día de Muertos. These two lectures offer two different approaches to understanding the rapid rise in popularity of Santa Muerte devotion. Whereas Eduardo González Velázquez explores the role of the cult with migrants and other marginalized sectors of Central American society, Eduardo García-Villada examines prayers to Santa Muerte for what they reveal about the attitude of her devotees to traditional power structures. To these two perspectives, The Comparison Project's director, Tim Knepper, offers brief reflections about the importance of Santa Muerte devotion for the theme of the lecture series: medicalized dying. In short, Santa Muerte protects and provides for her devotees in a way that the state and church cannot; her cult therefore serves an end not unlike that of bioethics: the attempt to control death.

Part II ("Religion and Medicalization") begins with Lucy Bregman's "Christians Encounter Death: The Tradition's Ambivalent Legacies," which argues that in spite of the diversity of practice and experience in the Christian tradition, the Christian understanding of death remains centered on the death and resurrection of Jesus Christ. One implication of this understanding, according to Bregman, is that given the violence, destruction, and link to sin involved in Jesus' death, it is difficult for Christians to consider death as "a natural event." Drawing on theology, art, and literature from the history of Christianity, Bregman shows how Christians instead understood there to be "something un-good about death as humans experience it" in that it can never be "completely erased or obliterated." Bregman then considers some recent challenges to and defenses of the traditional Christian understanding of death: mind/body dualism, death as a punishment for sin, the de-emphasis of bereavement, and the necessity of hell. Finally, Bregman contends that Christians should not "secede territory" to medicalized dying and the myth of "Progress" that undergirds it; rather, they should demonstrate its limitations and take courage in the Christian tradition's rich, counter-cultural understanding of death.

Christopher Key Chapple's "A Jain Ethic for the End of Life" examines the Jain practice of *sallekhana*, a "fasting unto death" by which some Jains die a conscious and peaceful death that expresses their core faith and ensures a good rebirth. Chapple explains that *sallekhana* is not for all Jains but rather only those who possess terminal diseases or are otherwise close to death, who have gone through a rigorous period of internal reflection, and who have received approval from the community. Nevertheless, as Chapple details, *sallekhana* has recently been challenged in the courts as a form of suicide. In 2015, the Indian state of Rajasthan banned *sallekhana* and deemed criminal anyone who would abet the practice. Although the Supreme Court of India put a stay on the ban temporarily, the case demonstrates yet another way in which traditional theologies of death and practices of dying have been challenged by modern, medicalized and legalized dying. Chapple, however, finds "instructive wisdom" in the Jain fast to death, which embraces death (near or at the end of life) rather than seeking to avoid it.

The third essay of Part II—Herbert Moyo's "The Ritualization of Death and Dying: The Journey from the Living Living to the Living Dead in African Religions"—focuses on the Ndebele people of Matabo in Zimbabwe, exploring how dying for them is a transition from the world of the living to the world of the living dead. Moyo discusses how death is a calling by the ancestors to join them. When this calling occurs, there is nothing that modern medicine can do to prevent it. However, since the journey to the world of the living dead is fraught with obstacles, it is necessary for the living to perform rituals for the recently deceased. Absent the correct performance of such rituals, ancestors will come back to trouble their relatives. In some cases, these traditional rituals are complicated by modern hospitalization and technology. This is especially the case with respect to the ritual that collects the soul of the recently deceased from the place of death. Nevertheless, argues Moyo, the Ndebele world continues to be one that ritualizes death rather than medicalizing death.

The last essay of Part II—Alyson Prude's "Death in Tibetan Buddhism"—explores Tibetan Buddhist understandings of the dying process and transition to the next life. Prude first describes end-of-life rituals and funerary customs, notions of timely versus untimely death, and the possibility of returning from death to one's previous life. In doing so, she maintains that such concepts and practices challenge reductionist western medical theories that refuse to allow for the continuation of consciousness at death. Prude then turns to the topic of medicalized dying, showing how the Buddhist understandings of karma and transmigration often mitigate against aggressive treatment options near the end of life that prolong life at all costs. For Buddhists in general, it is far more important to die peacefully, with clarity and acceptance of death. In Prude's words, "a good death is more valuable than a prolonged life."

Part III ("Religion and Bioethics") begins with Elliot Dorff's "Jewish Perspectives on End-of-Life Decisions," which itself begins by examining fundamental Jewish beliefs related to health care, most notably that the body belongs to God, that human beings have both the permission and the obligation to heal, and that the physician holds authority in decisions about health care. After a brief consideration of advance directives, Dorff devotes considerable attention to several bioethical dilemmas: foregoing life-sustaining treatment, artificial nutrition and hydration, pain control and palliative care, and medical experimentation and research. In many of these cases Dorff argues from the Jewish tradition for positions that seek to save lives and cure illness, while at the same time recognizing the limitations of medicine and the realities of dying. In particular, Dorff maintains that it is morally permissible to withhold or withdraw not only medicines and machines but also artificial nutrition and hydration in cases where patients are diagnosed with terminal, incurable illness. Dorff then ends his essay by examining care of the deceased, including Jewish norms about burial, cremation, autopsies, organ and tissue donation, and donating one's body to science.

Damien Keown's "Buddhism and Brain Death" next examines the definition of death as brain death, particularly as tensions exist between it and both classical Buddhist teachings about death and contemporary Buddhist attitudes toward death. After a brief introduction to Buddhist teachings relevant to death and dying, Keown considers some recent challenges to the contemporary brain-death definition of death as well as some proposals that human death should be redefined as the irreversible loss of consciousness. Keown then examines the compatibility of brain death with the classical Buddhist understanding of death, suggesting that Buddhism does not support the notion of cognitive death due to its rejection of mind-body dualism.[4] Finally, Keown explores contemporary attitudes toward brain death in two Asian countries that are heavily affected by Buddhism: Japan and Thailand. In neither case, according to Keown, is the concept of brain death well received at a popular level.

The third essay of Part III—Gerald Magill's "Ethical Engagement with the Medicalization of Death in the Catholic Tradition"—begins by summarizing some of the theological roots and tools of bioethical reasoning in the Catholic tradition, especially Natural Law and the Principle of Double Effect. Magill turns next to the teachings of the U.S. Catholic Bishops on issues related to health care, as articulated in the *Ethical & Religious Directives for Catholic Health Care Services*. Here, Magill underscores three points: first, extraordinary means of treatment are morally optional for patients; second, the patient plays a crucial, determining role in ascertaining whether a means of treatment is obligatory (ordinary) or optional (extraordinary); third, burden and expense are relative factors in this decision. Finally, Magill shows how these principles, tools, and insights apply in the case of three

[4]As Keown indicates, this view constitutes a reversal of the position that he espoused over 20 years ago in *Buddhism and Bioethics*, wherein he asserted that the concept of brain death would be acceptable to Buddhism.

bioethical issues that ethically engage issues surrounding the medicalization of death: maternal-fetal conflicts, patients approaching the end of life, and after-death dilemmas. Magill concludes that the flexibility and range of Catholic morality can offer astute guidance not only to those who follow this tradition but also to those with different faith perspectives (or none) when they encounter the heart-wrenching dilemmas that medical technology presents around death and dying.

Part III ends with Aasim I. Padela, and Omar Qureshi's "Islamic Perspectives on Clinical Intervention Near the End of Life," which argues that traditional Sunni Islamic ethico-legal views, in conjunction with the Islamic theological concepts of human dignity (*karāmah*) and inviolability (*ḥurmah*), provide the ethical grounds for non-intervention at the end of life. Padela and Qureshi begin with an extended exploration of the scriptural basis and legal rulings pertaining to end-of-life are in the four Sunni schools of law: Ḥanafī, Mālikī, Shafiʿī, and Ḥanbalī. They conclude that, despite some interesting differences between these rulings in these schools of law, clinical intervention is not morally obligated for Muslims. The authors next examine the theological concepts of *karāmah* and *ḥurmah*, arguing that both suggest against disrupting the sanctity and inviolability of the body through clinical interventions near the end of life. Finally, Padela and Qureshi close by commenting on the urgent need for a theologically rooted, holistic bioethics of caring for the dying from an Islamic perspective in order to better serve Muslim physicians and patients.

1.2 Comparison

Our final essay—"Comparative Conclusions"—is in fact four mini-essays, each of which was delivered at the final event of The Comparison Project's 2015–2017 series. The authors of these essays are the designers of the 2015–2017 series: Timothy D Knepper, The Comparison Project's director; Mary Gottschalk, who assisted Tim with the public programming and curricular offerings on death and dying; Lucy Bregman, our guest scholar, comparativist, and editor for the series on death and dying; and Allen Zagoren, who not only helped organize the series but also provided its kick-off lecture.

Allen Zagoren's conclusion offers us the perspective of a practicing surgeon. Hauntingly, Zagoren wrestles with the physician's instinct to save life at all costs—to defeat death wherever possible—even while understanding that this instinct often comes into conflict with what is best for a person and a society. More broadly, Zagoren's conclusion confronts many of the dilemmas facing humanity as we continue to develop the bio-technological means of prolonging life to greater lengths, if not of eventually extending life indefinitely. What will this future mean for traditional theologies of death and rituals of dying? What will it entail for socio-economic realities? Can we have the courage to face these changes even while we acknowledge the inevitability of death? Zagoren asks the tough questions, refusing to insult us with easy answers.

Lucy Bregman's conclusion focuses attention on the range of relationships between religious traditions and modern-western medicalized views of death. Chiding those who assume a facile model of uniform linear progress away from religious traditions and toward a rational scientific outlook, Bregman shows how religions have sponsored western medicine, resisted pieces of it, and blended with much of it. Bregman then asserts that although all cultures have some role for healers, it is the role of doctor as medical researcher that has become most problematic, both in our own culture and elsewhere. In awareness of this fact, Bregman believes we can counter the dominance of medicalization with other ethical and spiritual visions of the human good.

Mary Gottschalk's "layperson's" conclusion takes up three key questions that emerged for her over the course of the 2015–2017 programming cycle: (1) Does the fact that we have the medical means to cure disease or prolong life mean that we should do it? (2) What are the guidelines for determining what should be done and when? (3) Who should make this decision and how? In the first case, Gottschalk notes some religious obstacles to prolonging life at all costs; in the second, some internal tensions within traditions with regard to determining bioethical guidelines; and in the third, even more tensions with regard to who decides and how. Gottschalk then ends by offering an insightful "concluding thought": although different religious or philosophical traditions often come to similar bioethical conclusions based on quite different reasoning, there are also significant differences of opinion within most faith traditions.

Timothy D Knepper's comparative conclusion ends the essay and volume by attempting to explain one part of Gottschalk's "concluding thought"—viz., the striking similarities between the bioethical positions of different religions. Before doing so, however, Knepper attempts to "think together" The Comparison Project's diverse programming on death and dying by categorizing religious traditions' responses to medicalization in particular and modernity in general into those of resistance, accommodation, and innovation. Knepper then offers a framework for understanding religious bioethical reasoning by looking at its three sources: those set by the realities of dying, those informed by cultures and traditions, and those given by moral intuitions about killing. Drawing on the cognitive scientific approach of Pascal Boyer, Knepper maintains that religious reasoning about end-of-life issues might only constitute *post hoc* justifications of decisions that are already made by our moral intuitions in view of the realities of dying. If so, this is one compelling reason why different religious traditions that argue from different theological principles come to similar conclusions about bioethical issues.

Reference

Engelhardt, H. Tristram, Jr. 1991. *Bioethics and secular humanism: The search for a common morality*. Eugene: Wipf & Stock.

Part I
Death and Religion

Chapter 2
Death in Ancient Chinese Thought: What Confucians and Daoists Can Teach Us About Living and Dying Well

Mark Berkson

Abstract The foundational texts of the classical period of Confucianism and Daoism contain virtually no discussion of post-death existence or the nature of the afterlife. At the same time, these traditions devote significant attention to the ways death and loss impact our lives. Confucian texts such as the *Analects of Confucius* and the *Xunzi*, as well as the distinctive, profoundly influential writings of the Daoist Zhuangzi, contain teachings and stories about people facing their own deaths and dealing with the deaths of others. Both traditions offer guidance on living well in the face of mortality without recourse to notions of post-death existence. They do this by cultivating and sustaining connections and identification with larger realities beyond the individual. The Confucians emphasize human relationships, the realm of the familial (including ancestral), social, and political. Zhuangzi emphasizes the connection with the larger natural world, whereby we see the cycles of our own lives as part of the larger cycles of nature, and he points toward a path to losing one's self through contemplative and skillful practices, ultimately leading to a sense of equanimity in the face of death. I argue that underlying the differences between Confucian and Daoist forms of connection and consolation is the contrast between the Confucian notion of a narratively constructed self, an achievement of memory and ritual, and Zhuangzi's deconstruction of the narrative temporality that grounds Confucian notions of selfhood.

2.1 The "This-Worldly" Emphasis of Classical Confucian and Daoist Thought

Throughout Chinese history, most people have believed in a wide variety of supernatural beings, often grouped into the categories of gods, ghosts, and ancestors. The Chinese religious landscape has featured many different beliefs about, and practices oriented toward, an afterlife. And yet the most well-known, influential thinkers and

M. Berkson (✉)
Hamline University, St. Paul, MN, USA
e-mail: mberkson@hamline.edu

T. D. Knepper et al. (eds.), *Death and Dying*, Comparative Philosophy of Religion 2, https://doi.org/10.1007/978-3-030-19300-3_2

texts of China's two most important indigenous religious and philosophical traditions—Confucianism and Daoism—rarely address issues relating to an afterlife. In the rare cases that they do, they exhibit no certainty about its nature and usually direct our attention back to how we live our lives in this world. It is worth looking closely at how the Chinese thinkers find meaning and consolation in the face of finitude without recourse to personal survival of death. It is also instructive to explore the significant differences between the Confucian and Daoist perspectives and reflect on what each can reveal about the nature of mortal existence.

The fact that the worldviews and value systems of the ancient Confucian and Daoist texts do not rely on any kind of supernaturalism makes the perspectives of classical Confucian and Daoist thought useful for those looking for systems of meaning that are not dependent on any particular conception of a deity, supernatural beings, or an afterlife. It also makes it possible for people living millennia later in a radically different culture (e.g., the contemporary West) to incorporate Confucian and Daoist insights into our lives, whether we have existing religious commitments, are atheists, or are a part of the fastest growing "religious" demographic in America, the "unaffiliated" (including those who are "spiritual but not religious"). In many ways, these thinkers offer a critique of an overly medicalized approach that looks at the technology-enabled prolongation of life as a central goal and sees dying as a failure. The Confucians help us find meaning within the pain of grief and loss, more deeply understand the power and efficacy of death rituals, and more fully appreciate the nature of our relationships with the dead. The Daoist thinker Zhuangzi challenges us with the thought-provoking possibility that we can transform ourselves in such a way so as to avoid grief and accept mortality with equanimity and humor. And thinkers in both traditions present visions of what constitutes a good death.

I will focus on two of the most influential thinkers who lived during the "Golden Age" of Chinese thought, the Confucian Xunzi (ca. 312–230 BCE) and the Daoist Zhuangzi (369?–286? BCE). Their views represent key elements of Confucian and Daoist thought, although Zhuangzi's vision is so idiosyncratic that it is difficult to categorize him. I will also briefly address how Confucius himself addressed death.

The significant differences between these Chinese traditions and the Abrahamic, monotheistic traditions, as well as the differences between the two Chinese traditions themselves, make these thinkers valuable participants in our conversation about death and dying.[1]

2.2 Confucius

Unlike most religious texts, the *Analects* of Confucius indicates virtually no concern with the existence of an afterlife, and no metaphysical reflection on the nature of death. In one famous exchange, a disciple approached Confucius and said, "May

[1] Portions of this chapter appear in earlier publications. Some material on Xunzi first appeared in Berkson (2016). Some material on Zhuangzi first appeared in Berkson (2011).

I ask about death?" Confucius replied, "You do not understand even life. How can you understand death?" (11.12).[2] In one sense, Confucius seems to be saying that our task as human beings is to live well, to learn what it means to be a good human being, and therefore that we should not waste our time thinking about death.

At the same time, Confucius *does* talk about death frequently in the *Analects,* with many passages about grief and mourning, as well as examples of people on their deathbeds. So as we look again at Confucius' response, we can see an additional meaning. Only when one truly understands *life*—characterized by grief, loss, and one's own mortality—will one understand the meaning of *death*.

Confucianism as a tradition focuses on self-cultivation through learning, ritual, the arts, and physical practices. In contrast to a common view of the self in the West as a separate, rational, autonomous agent, the Confucian view is that the self is constituted by a nexus of relationships. We are who we are because of whom we have lived with, loved, and learned from. If you ask me who I am, my Confucian answer would be straightforward—I am a son, father, brother, husband, teacher, friend, and so on. To be a good person is to learn what it means to be good in each of these roles. To understand the Confucian approach to death is to keep in mind the importance of human relationships above all else.

For Confucius, feelings of grief are among the most important human emotions. They indicate, in a way nothing else can, the powerful connection we feel with other human beings. Thus, nothing could be a finer expression of humanity then the proper and complete expression of these feelings in the ritual act of mourning, especially when the deceased is a close family member. One of Confucius' disciples said, "I have heard the Master say that on no occasion does a man realize himself to the full, though, when pressed, he said that mourning for one's parents may be an exception" (19.17). Ritual in general occupies the highest place within Confucian thought, and death rituals are the highest form of all rituals.[3]

Confucians are known (and often stereotyped) for emphasizing formality and propriety. Yet Confucius explicitly tells his disciples that during mourning, feelings of grief take precedence over all other concerns. He advises, "With the rites, it is better to err on the side of frugality than on the side of extravagance; in mourning, it is better to err on the side of grief than of formality" (3.4).[4]

A notable feature of the *Analects* is that it highlights Confucius' response to the mourning of others. In a number of passages, Confucius' behavior shows that one's character is reflected not only in how one mourns, but also in how one responds to others who mourn. Confucius' expression, posture and behavior exhibit solemnity

[2] Citations from the *Analects* and the *Mencius* refer to D.C. Lau's translations (Confucius 1979; Mencius 1970). For the sake of consistency, all Chinese terms in this chapter are presented in Pinyin Romanization, even when the original translation uses Wade-Giles.

[3] For thoughtful discussions of death in Confucian thought, see Yearley (1995) and Ivanhoe (2003).

[4] We see examples of Confucius himself weeping and lamenting the deaths of his beloved friends and disciples to the point where he is criticized for going beyond propriety. When his favorite disciple Yan Yuan died, his followers said, "You are showing undue sorrow." Confucius replied, "Am I? Yet if not for him, for whom should I show undue sorrow?" (11.10).

and deep regard in the presence of one who has experienced loss. He will not enjoy himself or be casual around someone in the midst of grief, and he intentionally deprives himself of food, empathetically suffering with the mourner.[5]

In both the *Analects* and the *Zhuangzi*, the masters' deathbed experiences are described. The way that each faced death is an expression of the values that underlie their worldviews. First, we will look at Confucius' deathbed story:

> The Master was seriously ill. Zi Lu told his disciples to act as retainers. During a period when his condition had improved, the Master said, "(Zi Lu) has long been practicing deception. In pretending that I had retainers when I had none, who would we be deceiving? Would we be deceiving Heaven? Moreover, would I not rather die in your hands, my friends, than in the hands of retainers? And even if I were not given an elaborate funeral, it is not as if I was dying by the wayside." (9.12)

Confucius, no longer employed in any official capacity, is not entitled to retainers. Yet his disciples, out of respect and reverence, want to act as such. Confucius emphasizes that such deception in ritual practice is an offense and sharply criticizes officials for arrogating to themselves privileges to which they are not entitled. Even in the end, propriety must be upheld. Violating the rituals when conducting Confucius' own funeral would be a betrayal of the Way that Confucius held dearer than life itself.

Beyond this, however, Confucius emphasizes the importance of being surrounded by one's friends at death rather than having an elaborate funeral. Despite all of the emphasis placed in the *Analects* on formalities, at this crucial moment in one's life, the focus, in death as in life, is on our genuine connection with other human beings.

2.3 Xunzi

2.3.1 Xunzi's Defense of Confucianism

Xunzi, the first systematic Confucian philosopher, took up the task of defending Confucianism from philosophical attacks. At the same time, as someone who might be considered both a naturalist and a rationalist, Xunzi was disturbed by the beliefs in ghosts and spirits that motivated many individuals' participation in ritual, believing them to be not only misguided but also potentially harmful. He had to argue for the necessity of ritual activity for the individual and society *and* to reinterpret the way rituals were understood, so as to make participation compelling during a period of significant changes in and diversity of worldviews. Xunzi wanted to underscore the importance of rituals (especially death rituals) while interpreting them in a way that is entirely naturalistic, psychologically sensitive and profound, and deeply spiritual.

---

[5] See, for example, 7.9 and 10.25.

2.3.2 *Grief and Mourning*

In his detailed and moving account of the Confucian death rituals, Xunzi recognizes that the powerful feelings of grief that normally accompany the death of a loved one can become harmful or pathological if they are not properly expressed:

> Everyone is at times visited by sudden feelings of depression and melancholy longing.... A filial son who has lost a parent, even when he is enjoying himself among congenial company, will be overcome by such feelings. If they come to him and he is greatly moved, but does nothing to give them expression, then his emotions of remembrance and longing will be frustrated and unfulfilled, and he will feel a sense of deficiency in his ritual behavior. (Xunzi 1963, p. 109)[6]

The human response to the death of a loved one involves both strong emotions and changes in behavior. Paul Rosenblatt writes, "If people who are bereaved are to return to reasonably normal patterns of productivity and social life, they need to 'work through' the loss. Working through processes include acceptance of the loss, extinction of no longer adaptive behavioral dispositions ... and dissipation of guilt, anger and other disruptive emotions" (Rosenblatt et al. 1976, p. 6).

The Confucian rituals, Xunzi explains, provide an outlet for that grief, taking the raw emotion and cultivating a beautiful form of spiritual communion with the deceased that brings higher goods to the mourner: "The sacrificial rites originate in the emotions of remembrance and longing, express the highest degree of loyalty, love, and reverence, and embody what is finest in ritual conduct" (Xunzi 1963, p. 110).[7] As we will see, they also provide for a gradual return to normal life, as the transition period is difficult and fraught with danger to the psyche.

2.3.3 *Ambivalence and Emotional Conflict*

Death rituals must help mourners deal with conflicting emotions and ambivalent attitudes. Confucian death rituals provide a moving example of the maintenance of important contradictions for the sake of both stability and meaning. Xunzi recognizes that the mourner is in a state of conflict, because she knows that the loved one is gone, but cannot come to accept that fact (cognitively, emotionally, or behaviorally). The death ritual allows for this by treating the dead "as though" alive: "In the funeral rites, one adorns the dead as though they were still living, and sends them to the grave with forms symbolic of life. They are treated as though dead, and yet as though still alive, as though gone, and yet as though still present. Beginning and end are thereby unified" (Xunzi 1963, p. 103). There is no naive self-deception here;

[6] Unless otherwise noted, all Xunzi translations are taken from the translation by Burton Watson (1963).

[7] Psychologist Rosemary Gordon writes, "If the rites are successful, they provide a socially acceptable expression for grief and at the same time, by acting as a framework, they help a person to resist the disintegrating force of this grief" (Gordon 1978, p. 78).

rather, the mourner is engaging in a sophisticated form of pretending that creates a healthy outlet for the emotions.

The challenges of ambivalence can be addressed effectively in the symbolic and aesthetic realms of ritual, which are able to sustain and often reconcile deep psychological tensions in ways that logic and discursive reason cannot. In Xunzi's description of death rituals, an object can both be present and absent; an act can be "scripted" yet "spontaneous"; one can be treated "as if alive" and "as if dead." Xunzi writes, "Beauty and ugliness, music and weeping, joy and sorrow are opposites, and yet rites make use of them all, bringing forth and employing each in its turn" (Xunzi 1963, p. 100).[8]

One of the ways that Confucian ritual allows for the proper expression of these contradictory thoughts and emotions is through the inclusion of altered objects in the coffin. Xunzi writes:

> Articles that had belonged to the dead when he was living are gathered together and taken to the grave with him, symbolizing that he has not changed his dwelling. But only token articles are taken, not all that he used, and though they have their regular shape, they are rendered unusable.... (Xunzi 1963, p. 104)

This is known in anthropological terms as a "tie-breaking ritual." The strong ties to the deceased make reentry into everyday life and the learning of new behaviors difficult. Thus, ritually marking the end of the relationship (and the beginning of a new one based solely on memory) becomes important. There are many types of tie-breaking customs, including disposal of personal property of the deceased and taboos on the name of the deceased. In modern Taiwan, there are rituals in which each family member "takes hold with one hand of a long, hempen rope, one end of which is attached to the sleeve of the dead person." Each portion of the rope is then cut, "whereupon each person wraps up his portion in a sheet of silver Hades money and burns it. This is called 'cutting off lots'" (Paper and Thompson 1998, p. 48).

Anthropologist Roy Rappaport states that Arnold Van Gennep "observed that particular kinds of rituals tend to include physical acts that seem to be formally similar to whatever they seek to accomplish. Rites of separation, for instance, often include such acts as cutting something, perhaps the hair" (Rappaport 1999, p. 141). This is making material and observable what is non-material. The materiality of the display, Rappaport suggests, makes it "performatively stronger" than merely saying words to the effect that the deceased is gone forever. This greater investment makes it more likely that the survivor will accept the death and begin to develop emotions and behaviors better suited to the transformed relationship.

[8] The tension and ambivalence experienced during the mourning process was a significant theme in the work of Freud, who cites a number of cases of patients refusing to accept the death of a loved one. He describes the process whereby the ego splits, with one part accepting death, and the other denying it. Freud describes a child's reaction to his father's sudden death: "I know father's dead, but what I can't understand is why he doesn't come home to supper" (Freud 1989, p. 92). Likewise, Xunzi writes about the filial son who observes his lord or parent no longer breathing: "Weeping and trembling, he still cannot stop hoping that the dead will somehow come back to life; he has not yet ceased to treat the dead man as living" (Xunzi 1963, p. 99).

2.3.4 Facilitating Transition

It has been said that the mourner has two powerful conflicting desires to combat at a funeral: the desire to jump into the grave with the deceased (the inability to let go) and the desire to run away (the revulsion at the corpse). Xunzi believed that death rituals must account for both, allowing for their proper (and non-pathological) expression and harmonization. Robert Jay Lifton writes of the "universal dilemma around ties to the dead: on the one hand, the survivor's need to embrace them, pay homage to them, and join in various rituals to perpetuate the relationship; on the other the tendency to push them away, to consider them tainted and unclean, dangerous and threatening" (Lifton 1983, p. 92).

Revulsion at the sight of the corpse must be addressed by a funeral ceremony that accounts for both the passage of the deceased from taboo corpse to revered ancestor and the passage of the survivor from grieving mourner to nonmourner. Xunzi is doing just this when he writes:

> It is custom in all mourning rites to keep changing and adorning the appearance of the dead person, to keep moving him farther and farther away, and as time passes, to return gradually to one's regular way of life. It is the way with the dead that, if they are not adorned, they become ugly, and if they become ugly, then one will feel no grief for them. Similarly, if they are kept too close by, one becomes contemptuous of their presence.... The gentleman is ashamed to have such a thing happen, and therefore he adorns the dead in order to maintain the proper reverence, and in time returns to his regular way of life in order to look after the wants of the living. (Xunzi 1963, pp. 99–100)

Lifton writes, "What is involved is the symbolic transformation of a threatening, inert image (of the corpse) into a vital image of eternal continuity (the soul)—or of death as absolute severance to death as an aspect of continuous life." This is because, he goes on, "Survivors' psychological needs include *both* connection and separation" (Lifton 1983, p. 95). Xunzi allows for a gradual separation and a return to everyday life. This way, the necessary severance can occur within what Lifton calls "a sustaining matrix of connection."[9]

For Xunzi, the process of working through loss and reintegrating back into daily life can only be accomplished by ritual, which allows for the step-by-step return to reality that is the balance between a jarring, sudden resumption of normal activity and the lingering inability to adjust to the new reality. Xunzi observes:

> When a wound is deep, it takes many days to heal; where there is great pain, the recovery is slow.... The (mourning period) comes to an end with the twenty-fifth month. At that time, the grief and pain have not yet come to an end, and one still thinks of the dead with longing, but ritual decreed that the mourning shall end at this point. Is it not because the attendance on the dead must sometime come to an end, and the moment has arrived to return to one's daily life? (Xunzi 1963, p. 106)

[9] There are a variety of pathologies that might emerge in the aftermath of the death of a loved one. One is known as "morbid grief reaction syndrome," wherein mourning is delayed for a long period of time, months or even years after a loved one dies. Another is "pathological mourning" or "prolonged grief disorder," which occurs when the grieving person is unable to let go of the deceased, attempting to preserve as exactly as possible the objects or residence of the deceased and to continue, without alteration, the behavior and activities that the mourner carried out before the death.

Ultimately, Xunzi contends, the period of bereavement must be ritually terminated. This does not mean that the pain will be gone, but that one must force oneself to reestablish a routine even when one does not want to. This required "return to the everyday" recognizes the danger of an ever-deepening spiral of sadness and pain after the death of a loved one. As we have seen, the rituals allow these feelings, in all of their power, to be given time in which to be expressed while recognizing that reentering one's everyday life routine is an essential part of the healing process.[10]

2.3.5 The "As If" Attitude

Xunzi frequently employs the term *ru*, "as if"—in fact, we might say that it is one of his most important hermeneutic contributions. Here is one representative passage where Xunzi, using the character *ru* (translated "as though"), shifts the interpretation of a ritual from a literal, supernatural one to a symbolic one:

> When conducting a sacrifice, … one speaks to the invocator *as though* the spirit of the dead were really going to partake of the sacrifice. One takes up each of the offerings and presents them *as though* the spirit were really going to taste them…. When the guests leave, the sacrificer … weeps *as though* the spirit had really departed along with them. How full of grief it is, how reverent! One serves the dead *as though* they were living, the departed *as though* present, giving body to the bodiless and thus fulfilling the proper form of ceremony. (Xunzi 1963, pp. 110–111; italics mine)

Xunzi's reinterpretation and defense of ritual addresses two problems. He is concerned with those for whom the symbols have lost their power. He is also worried about people reading the objects literally rather than symbolically, which would lead to delusion (e.g., believing in the existence of ghosts, spirits, and a caring, intervening heaven).

Instead of offering a picture of a benevolent deity and immortal soul to provide solace to the mourner, Xunzi elucidates the way the symbolizing attitude can help the mourner through the difficult transition. Xunzi argues that it is one of the unique benefits of ritual that it can embrace and harmonize opposites. Gordon observes: "The ability to symbolize, which is the most essential ingredient in all creative endeavor, rests very firmly indeed on the capacity to accept both … the forces of life and the forces of death…. in short, it involves the coming to terms with paradox (Gordon 1978, p. 24). This ritual participation can be seen, perhaps, as a form of

[10]We can also imagine cases where someone is emotionally ready to reenter the everyday world but is worried about appearances. For this reason as well, there is a ritually approved moment, often represented in contemporary Chinese culture by the second or final burial, in which one is "allowed" to terminate the formal mourning process. Death rituals involve two kinds of passages—the deceased moves from the realm of the living through a liminal state to the realm of the dead, a status change that must be culturally marked and accepted (there may be further passages at later times for the deceased, but this first one is fundamental), and at the same time, loved ones move from non-mourners to mourners to non-mourners once again. The second or "final" burial can be seen as a ritual that can mark the official end to the bereavement period, the final acceptance of the loss, and a commitment to terminate mourning.

sophisticated pretending or play, artificial and invented, yet necessary for conflicted and fragile beings such as ourselves.

2.3.6 *Symbolic Realism*

Xunzi can be understood as a "symbolic realist," a term borrowed from Robert Bellah. Bellah is a thinker whose task is, in many ways, consonant with that of Xunzi—offering a reinterpretation of religious notions that conveys the powerful truths of human life, that is not in contradiction with natural or social science, and that can appeal to intellectuals faced with the choice between abandoning religion and reinterpreting it. Here is one definition Bellah provides: "Symbolic realism simply holds that religious symbols are not primarily social or psychological projection systems (though they contain some projective elements), but the ways in which persons and societies express their sense of the fundamental nature of reality, the totality of experience" (Bellah 1970, p. 113).

Xunzi frequently speaks explicitly of the symbolic function (*xiang*) of ritual and music. When discussing death rituals, for instance, he writes, "Hence the three months of preparation for burial symbolizes that one wishes to provide for the dead as one would for the living" (Xunzi 1963, p. 99). He also discusses the symbolic role of such ritual accoutrements as "the mourning garments and cane, the mourning hut and gruel, the mat of twigs and the pillow of earth" (Xunzi 1963, p. 110).

For Xunzi, the truth lies not behind the symbols, but *in* them. The symbols do not hide or distort reality; they are the only way to fully express, apprehend, or experience reality. The symbolic realist tells us that those things that are manifested in and shaped through ritual symbols—non-material, vitally important aspects of human experience such as reverence, benevolence, grief, loyalty, and connectedness—are *real* and cannot be reduced to some other language or explained away.

The role of the "as if" attitude and the centrality of creative symbolization in Xunzi's thought mean that, in a way, ritual for Xunzi is like art or literature. One is employing a similar attitude to that which one has when reading a novel or listening to music: I know that what I am experiencing is not the product of a beautiful, harmonious world that cares for our interests; nevertheless, what has been *created* is beautiful and harmonious, and makes living in the world both possible and meaningful. Fiction, like ritual, can be one of the most powerful avenues directly into reality. However, the benefit only comes if the reader "buys into the illusion" or takes on the "as if" attitude. Both fiction and Xunzian ritual (which can be seen as a beautiful, necessary fiction) can act to reveal the truth rather than conceal or flee from it. The participant in ritual who gains a deeper understanding of and connection with reality is analogous to the reader who benefits through immersion in a work of fiction; neither denies the invented nature of what transforms them.

Xunzi's point is that it is not the act of invention that is pathological, but the fleeing from reality and the embrace of supernaturalism that so often accompany it. For Xunzi, the religious person and the artist can be both honest (about themselves and the world) and well-adjusted, both realistic/scientific *and* creative. Ritual and art can be forms of uncovering reality, not merely evading it.

2.3.7 Xunzi's Contemporary Relevance

The awareness of fabrication in rituals can be quite threatening when it undermines the foundations given to ritual in most traditions, such as divine origin. Rappaport highlights the dilemma for many moderns: "The epistemologies that have been spectacularly successful in illuminating the ways in which physical aspects of the world work, when shone on humanity's conventional foundations, show them to be fabrications and thus, in a world in which objectivity and fact seem to own truth, delusory" (Rappaport 1999, p. 451). In the face of this, one can either deny the methodology that leads to this conclusion (and thus the fact of "fabrication"), which is the tactic of fundamentalists when it comes to the findings of natural or social science; or, one can accept the findings of these methodologies and risk alienation from religion and ritual. However, when Xunzi illuminates the fact of fabrication, he does so in a way that elicits not disillusionment or rejection, but rather reverence and admiration for the human achievement. One might say that in the case of ritual, the "magic" is even *more* powerful when one knows how the trick works. Xunzi reveres the tradition not *despite* its fabricated nature, but precisely *because* of it.

Xunzi's interpretation of rituals—and death rituals in particular—is designed to avoid falling prey to supernaturalism, irrationalism, wish-fulfillment, and literalism, on the one hand, and reductive rationalism, desanctification, and alienation, on the other hand. He shows us the possibility of a tradition that recognizes the connections that constitute and sustain us while unflinchingly facing the reality of finitude, death, and a world that is majestic but indifferent. What is remarkable about Xunzi's Confucianism—and what separates it from the vast majority of other traditions—is that it does not assume that there is an underlying order (*nomos*), a primordial, divinely sanctioned harmony or goodness to the natural world or cosmos. (This helps Xunzi avoid the problem of theodicy that most religious visions face, especially when dealing with death.) For Xunzi, ritual does not sanctify a pre-existing, discovered order; ritual itself creates this order even as it sacralizes it. One celebrates not only the order itself but also the ongoing creating and sustaining of it in ritual activity. This allows for the maintaining of relationships with the dead and sustains meaning in the face of loss.

2.4 Zhuangzi

2.4.1 The Challenge of Zhuangzi's Vision

It is no exaggeration to say that the fourth century BCE Chinese thinker Zhuangzi possessed one of the most distinctive voices of any writer in history and was one of those rare individuals whose radical, provocative vision causes us to reevaluate our most fundamental beliefs and values.[11] Certain elements of his thought, especially those regarding death, test the limits of the recognizably human.

The book that bears his name is a challenging text filled with fantastic tales, parody, hyperbole, paradox, riddles, and humor. Because Zhuangzi is a language skeptic and a perspectivist, he is often maddeningly difficult to pin down. But his playful slipperiness is driven by a fairly coherent and often compelling philosophical picture of the way human beings and the world are and, in light of these facts, the best way for us to live and die. Many stories within the text depict people facing their own imminent deaths or the death of a loved one; other characters muse about death and the proper attitude toward it. Coming to terms with change and mortality is a central aspect of Zhuangzi's vision. Zhuangzi believes that most, if not all, human suffering is self-inflicted, and that changing the way we look at the world allows us to move "freely and easily" through the world. He recognizes that both the prospect of our own deaths and those of loved ones produce fear and grief that, in his view, lead to a great amount of avoidable human suffering. Zhuangzi aims to free human beings from those habits of mind that obstruct our nature and undermine our ability to live with ease.

2.4.2 *Skepticism Regarding Death*

One of the most prominent themes encountered in the text, which is explicitly applied on a number of occasions to the problem of facing death, is skepticism. Zhuangzi applies his skepticism to language (the ability of words to convey stable meaning, represent reality, etc.), senses (the reliability of our senses, and even our ability to know wakefulness from dreaming), and ethics (our ability to know right from wrong). As Paul Kjellberg, Philip J. Ivanhoe, and Lisa Raphaels have pointed out, Zhuangzi's skepticism is not debilitating; it is not like that of the ancient Greek skeptic Cratylus, who is paralyzed by not knowing since he cannot find any basis on which to act (or not act). Zhuangzi's skepticism instead aims to break us of certain habits, such as acting on the false certitudes of senses, language, and judgments.[12] Zhuangzi applies his skepticism to our common assumptions about life and death:

> How do I know that loving life is not a delusion? How do I know that in hating death I am not like a man who, having left home in his youth, has forgotten the way back? .... How do I know that the dead do not wonder why they ever longed for life? (W, pp. 42–43)

Zhuangzi's skepticism reminds us that we simply do not, and cannot, know what happens after death. He plays on our lack of knowledge by suggesting that, for all

---

[11] Given the complexities surrounding issues of authorship in *Zhuangzi*, I consider a view "Zhuangzian" if it is found in the Inner Chapters, and passages from elsewhere are used only if they are consistent with this view. Page numbers for *Zhuangzi* citations refer to the Victor Mair translation (Zhuangzi 1994) unless otherwise noted. When A.C. Graham's translation (Zhuangzi 1989) is used, a "G" will precede the page number; if Burton Watson's translation (Zhuangzi 1964) is used, a "W" will precede the page number.

[12] See Kjellberg and Ivanhoe (1996).

we know, death could be a far superior state. After reminding us not to assume that death brings about an unfortunate state, Zhuangzi goes on to muse that since death will bring about the end to the problems that plague us in life, death can be seen as a relief. Death is often characterized in the text as a rest following the frenzied activity of life or a relief of life's accumulated tensions and anxieties, as when Zhuangzi says, "The Great Clod burdens me with form, toils me through life, eases me in old age, rests me in death" (p. 59). Zhuangzi's sages think of life as "an obstinate wart or a dangling wen, of death as bursting the boil or letting the pus" (G, p. 89).

2.4.3 *Accepting Death and Going Along with Change*

2.4.3.1 A World of Ceaseless Transformation

Zhuangzi places a high value on life, praising those who live out their full lifespans. This view exists alongside his philosophy of accepting death. These positions are reconciled by the fact that the life he values is characterized by ongoing transformation and understood as a manifestation of the natural dao, a way of life/death cycles. Zhuangzi believes not only that these two positions are not in tension, but also that one who values life must accept death as an inseparable dimension of the life he values. The dao, in both the *Dao De Jing* and the *Zhuangzi*, while it may be "eternal" (*chang dao*), is *not* static. It is a way of change, of ceaseless transformation. Zhuangzi vividly depicts the changing world:

> Decay, growth, fullness and emptiness end and then begin again…. The life of things is a gallop, a headlong dash—with every movement they alter, with every moment they shift. What should you do and what should you not do? Everything will change of itself, that is certain! (p. 103)

2.4.3.2 One's Own Death

Given that the world is characterized by continuous change, how should one best live? Zhuangzi advises us to "follow along" with change, preserve equanimity, and make the mind "free-flowing." Zhuangzi's sages do not become attached to the status quo or some future ideal, and they are not driven by plans. Instead, they act in accordance with life's ongoing transformations and thus are able to wander freely and easily in the world: "Your master happened to come because it was his time, and he happened to leave because things follow along. If you are content with the time and willing to follow along, then grief and joy have no way to enter in" (W, pp. 48–49). Elsewhere, Zhuangzi says:

> Life and death, preservation and loss, failure and success, poverty and wealth…. Day and night they alternate before us, but human knowledge is incapable of perceiving their source. Therefore, we should not let them disturb our equanimity, nor should we let them enter our numinous treasury. To make the mind placid and free-flowing without letting it be dissipated in gratification…. This is what I mean by wholeness of one's abilities. (pp. 47–48)

Zhuangzi observes that the continuous play of emotions, from sadness to joy to anger, is part of what it is to be human. However, we can nonetheless achieve equanimity within the flux of emotional and sensory experience. This does not demand removing oneself from society or closing off one's senses, for it is achieved in the midst of activity, what Zhuangzi terms "tranquility in turmoil" (Watson's translation is "peace in strife"; *ying ning*) (p. 57).

"Going along with" the transformation of things, at its most radical, involves accepting severe illness, mutilation, and death. Although one tries to avoid situations likely to lead to death, Zhuangzi's exemplars demonstrate the way to respond to life-threatening developments when they do, as they inevitably will, occur. In one passage, Master Yu suddenly falls ill and becomes deformed, yet remains "calm and unconcerned," expressing only wonder at the new shapes his body is taking. When asked if he resents these changes, he replies:

> Why no, what would I resent? If the process continues, perhaps in time it'll transform my left arm into a rooster. In that case I'll keep watch at night.... I received life because the time had come; I will lose it because the order of things passes on. Be content with this time and dwell in this order and then neither sorrow nor joy can touch you.... There are those who cannot free themselves because they are bound by things. (W, p. 81; see also [Mair,] p. 59 and p. 169)

One is tempted to say that Master Yu fell victim to a disfiguring disease. Yet this passage suggests that words like "disease" are mere designations, impositions upon the actual reality: Master Yu's body underwent transformations. He was not gripped with the fear that would follow from the designation of a disease. There is no pathos, no lamentation, no resistance.

Zhuangzi's sages exemplify acceptance of death, not grudgingly but with equanimity and good humor. There is no sense that they would want things to be otherwise. A.C. Graham writes that "for Zhuangzi, the ultimate test is to be able to look directly at the fact of one's own physical decomposition *without* horror, to accept one's dissolution as part of the universal process of transformation" (Graham 1989, p. 203). The acceptance of death, Zhuangzi shows us, depends upon our acceptance of the continuous transformations that we undergo moment by moment.[13]

2.4.3.3 Death of Others

Confucian thinkers insist that grief is the natural human response to the death of a loved one. For Confucians, we mourn because we love. How one grieves reflects both the nature of the relationship to the deceased and one's character.

Zhuangzi's perspective is profoundly different. For Zhuangzi, grief and sadness only show that one's mind is still prey to emotional storms that disturb equanimity and make true freedom impossible. He thus praises Mengsun Cai for avoiding grief, even over the loss of his mother:

[13] For a discussion of the radical implications of this view of accepting transformation, see Yearley (1983).

> When Mengsun Cai's mother died, … he did not grieve in his heart…. In his case, though something may startle his body, it won't injure his mind …. Mengsun alone has awoken …. What's more, we go around telling each other, I do this, I do that—but how do we know that this "I" we talk about has any "I" to it?… Be content to go along and forget about change and then you can enter the mysterious oneness of Heaven. (W, p. 85)

Zhuangzi draws an explicit connection between change and the lack of any stable identity or "I." Those who go along with change, rather than attempting to arrest it through the creation of selves or stable identities, experience the death of others with acceptance and tranquility. For Mengsun Cai, his mother's death was simply one more moment in the ongoing process of change; he can let go of her and therefore suffers no emotional pain. His mind cannot be injured because he has awakened to the nature of transience and its corollary, the absence of any enduring, substantial "self" or "I." Zhuangzi illustrates this point with a dramatic deathbed scene:

> Suddenly, Master Lai grew ill. Gasping and wheezing, he lay at the point of death. His wife and children gathered round in a circle and began to cry. Master Li, who had come to ask how he was, said, "Shoo! Get back! Don't disturb the process of change!" (p. 81)

The wife and children of the dying man respond in a way that Confucians would argue is perfectly natural—they cry. Master Li's response seems callous. Yet Master Li clearly cares about the dying Master Lai; after all, he had come to see how he was doing. His response can be seen as concern for Master Lai during his dying experience. Master Lai is going through a process of change, a process to which he must yield. His family, unable to accept this transformation, cries; they are unable to let go with equanimity, and therefore will be a disruptive element in Master Lai's death.

A famous passage in the Outer Chapters describes Zhuangzi's response to the death of his own wife to whom he was married for quite some time and with whom he raised children. His friend Huizi finds Zhuangzi pounding on a tub and singing, which Huizi finds disturbing. He asks Zhuangzi if that isn't going too far and Zhuangzi replies:

> Zhuangzi said, "You're wrong. When she first died, do you think I didn't grieve like anyone else? But I looked back to her beginning and the time before she was born…. In the midst of the jumble of wonder and mystery a change took place and she had a spirit. Another change and she had a body. Another change and she was born. Another change and she's dead. It's just like the progression of the four seasons, spring, summer, fall, winter…. If I were to follow after her bawling and sobbing, it would show that I don't understand anything about fate. So I stopped." (W, p. 113)

In this passage, we again see an emphasis on the need to accept the transformation of death just as one accepts all of the transformations that produce life. In the aftermath of the death, Zhuangzi employs a larger, "cosmic" perspective that sees the coming to be and passing away of his wife as manifestations of a larger, ongoing process of coming together and dissolution. Resisting or lamenting any transformation, Zhuangzi tells us, indicates a lack of understanding.

Another element present here is the reference to life and death as *natural* processes analogous to the changing of the seasons. It is not merely that the process of dying is another transformation in an endless series of changes, but that the cycle of life and death is *the* fundamental natural process. Zhuangzi's sages recognize that we are beings created and sustained by natural processes, and thus must yield to all natural transformations. The same forces that have brought one into being will lead, inevitably, to one's death. Accepting and affirming life requires the same attitude towards death.

2.4.3.4 Natural Cycles of Life and Death

The theme of cyclicality is found throughout Zhuangzi's writing: "Over and over turns the seamless cycle of beginning and ending" (pp. 60–61). As we accept the passing from one season to the next, Zhuangzi implies, so we must accept the life and death of a human being.

Seen in this context, death is understood as a completely natural phenomenon, a fundamental aspect of the dao: "Life and death are destined. Their constant alternation, like that of day and night, is natural/due to heaven" (p. 53). Zhuangzi reminds us that "the myriad things all come out from the wellsprings of nature and all reenter the wellsprings" (p. 173).

Seeing the life and death of an individual as fitting into the larger patterns of the dao suggests that the *kind* of change is significant, and that the treatment of natural deaths will differ from that of unnatural deaths. This is why there is no contradiction between Zhuangzi's ideal of living out one's natural life and avoiding putting oneself at risk, on the one hand, and his emphasis on accepting disease and death as simply continuing transformations, on the other. It is only *natural* deaths that are to be accepted, not those that are brought about by ambition or greed, which are the products of the mind's unnatural tendencies. The notion of *wu wei,* effortless (natural, spontaneous) action, involves acting in accordance with one's nature and the dao.[14] One should only die when it is one's time to die.

Many of the death-related themes encountered in the Inner Chapters appear in Zhuangzi's own "deathbed story," including the significance of nature, the acceptance of death, and the questioning of traditional norms and values:

> When Zhuangzi was on the verge of death, his disciples indicated that they wished to give him a sumptuous burial. Zhuangzi said, "I shall have heaven and earth for my inner and outer coffins, the sun and moon for my paired jades, the stars and constellations for my round and irregular pearls, and the myriad things for my mortuary gifts. Won't the preparations for my burial be quite adequate? What could be added to them?"
>
> "We are afraid that the crows and the kites will eat you, master," said the disciples.
>
> Zhuangzi said, "Above, I'd be eaten by the crows and the kites; below, I'd be eaten by mole crickets and ants. Why show your partiality by snatching me away from those and giving me to these?" (p. 332)

Zhuangzi playfully mocks those who put so much emphasis on proper burial accoutrements. For these, Zhuangzi substitutes all that surrounds him in the natural world. Zhuangzi focuses on the natural world and the heavens rather than the human beings around him. Humans are grouped along with all other things, among the "myriad creatures." Unlike Confucius on his deathbed, Zhuangzi does not mention the

[14] For an excellent discussion of *wu wei*, see Slingerland (2007).

importance of being surrounded by friends or disciples; rather, he emphasizes fitting into the natural world and the cosmos in a larger sense.

By asking, "What could you add?" Zhuangzi points to the enduring and pervasive human problem: the desire to meddle, to adorn and improve, a desire that is not merely unnecessary, but often harmful. For Zhuangzi, quite unlike the Confucians, dying needs no cultural embellishment.

2.4.3.5 The Experience of Unity

The *Zhuangzi* is filled with passages that suggest a related way to come to terms with death—undermine the notion that there is even a duality of "life" and "death." Zhuangzi calls into question the very concepts of beginning/end and before/after. By deconstructing the temporal distinctions that underlie our conceptions of an absolute beginning (birth) and an absolute ending (death), Zhuangzi points to a realm of no birth and no death—rather, there is simply an ongoing series of transformations.

Zhuangzi argues that absolute boundaries do not exist in nature. We mistakenly think they exist because we misunderstand how language works, e.g., thinking that a word like "birth" or "death" points to an absolute distinction. Zhuangzi points out, however, "The Way has never known boundaries" (W, p. 39). Because the dao has no absolute boundaries or divisions, the sage recognizes that "past," "present," and "future" are simply mental constructions: "Envisioning uniqueness, he could eliminate past and present. Eliminating past and present, he could enter the realm of lifelessness and deathlessness" (p. 57). In each moment, experienced just as it is without reference to past or future, one enters into a realm of "no dying, no being born" (*bu si bu sheng*) (p. 340).

The experience of the sage, then, is unity rather than division. There is the unity of life and death, and also the notion of unity with all things, which is why Zhuangzi's vision is sometimes described as "mystical." Since Zhuangzi's unity is with the dao and the natural world rather than with a deity or transcendent force, it can be seen as "nature mysticism."

In his vision, categorical separations are created only in the human mind through concepts. This is why Zhuangzi proclaims, "The perfect man has no self." Zhuangzi deconstructs binary oppositions (self/other, life/death) and gives examples of sages who experience unity (non-separation) with all other things:

> The spiritual man is of such integrity that he mingles with the myriad things and becomes one with them. (p. 7)

> Forget all relationships and things; join in the great commonality of boundlessness. (p. 99)

Thus, the sage realizes that life and death are interdependent aspects that are united within the dao.

2.4.3.6 No "Premature" Death

An implication of this recognition of unity is that "premature death," or any "untimely" death, is merely a construction based on a particular temporality—i.e., narrative temporality—that produces a "desirable" and "undesirable" time to die. In the realm of no-boundary, there is no ultimately good or bad time to die, no early or late, no premature or timely death. The entire notion of "a life" or "a lifespan" is a creation of the mind. Zhuangzi advises, "Forget how many years there are in a lifespan…. If you ramble in the realm of no-boundary, you will reside in the realm of no-boundary" (p. 23).

Once the sage is able to accept all transformations as they come, there is no conception of dying at the "wrong time" unless the death is due to human stupidity, in which case it is always the "wrong time," no matter how old the person is. As long as the death is the result of natural transformations, there can be no "right" or "wrong" time to die.[15] This is an instance of where Zhuangzi's thought differs dramatically from Confucius'. For Confucius, the death of a young person on the path of self-cultivation is fundamentally tragic. For Zhuangzi, the timing of death is irrelevant:

> The sage delights in early death, he delights in old age. (W, p. 77)

The profound differences between the Confucian and Zhuangzian notions of "premature death" are due to their differing conceptions of temporality. The Confucian notion of temporality sees the human life in terms of narratives. A life unfolds in stages, which are marked by rites of passage (e.g., the "capping ceremony," marriage, the 60th birthday). There are roles, duties and virtues appropriate to the different stages of life. The narrative conception is the framework against which Confucian notions of cultivation are developed.

For Confucians, if death is seen as something to be legitimately feared, it is not because of any existential terror, but rather because it represents a potential premature end to the narrative (and thus the loss of any possibility for further self-cultivation and development) and a loss of all possible goods. This is why Confucius says of his beloved disciple, Yan Hui, who died at a young age, "I watched him making progress, but I did not see him realize his capacity to the full. What a pity!" (9.21).

We can contrast the Confucian "narrative" with Zhuangzi's understanding of temporality. According to Zhuangzi, all narratives are based on constructions given to us by society—e.g., about the appropriate time to do certain things, about which roles should be occupied when, about what constitutes successful performance of our role-based obligations, etc. For Zhuangzi, these are constructed overlays on top of, and often obstructing, a ceaseless flow of life that can be experienced in its

[15] This sentiment is echoed by Cicero, who wrote, "Let us get rid of such old wives' tales as the one that tells us it is tragic to die before one's time. What 'time' is that, I would like to know? Nature is the one who has granted us the loan of our lives, without setting any schedule for repayment. What has one to complain of if she calls in the loan when she will?" (Potter 1988, p. 231). A contrary view is captured in the words of Zoe Akins, who observes in a Confucian fashion, "Nothing seems so tragic to one who is old as the death of one who is young, and this alone proves that life is a good thing" (Potter 1988, p. 227).

immediacy at any time. Without a narrative conception, there can be no sense of a life being cut short before the goals are reached or the story is completed. Given that life consists of moment-by-moment transformation, the best way to live does not involve cultivation or development; rather, Zhuangzi tells us to just "ride along with things" and let oneself "wander" rather than progress on a well-formulated path.

2.5 Comparative Reflections: Modes of Connection

Personal immortality, understood in a literal fashion, is not an option presented in the work of Confucius or Zhuangzi, or in the other major classical Confucian or Daoist texts. What both Confucians and Daoists show us is this: *We live and die well when connectedness and continuity are recognized, experienced, and appreciated through various practices.*

We can see that the Confucians and Zhuangzi emphasize what might be called different "modes of connectedness and continuity," ways in which aspects of who we are live on after our death. Confucius emphasized family and other human relationships, sustained by learning, ritual, and text, all of which are made possible through the act of *remembering*. We remember our dead in the hopes that we too will be remembered. Zhuangzi emphasized the absence of boundaries between us and nature. This is realized by *forgetting*—by letting go of social norms and expectations, and ultimately the very *self* that we construct.

Both Confucian and Zhuangzian thought teaches that overcoming separation and self-centeredness and making connections with larger realities are what provide meaning in life and solace in death. A number of types of connection and continuity are emphasized in Chinese thought[16]:

1. *Biological-familial*[17]: The Confucian emphasis on the family and lineage, and its elevation of filial piety with its demand not only to honor the parents but also to provide descendants, is a clear example of this recognition.[18]

[16] This list, and much of my understanding of the importance of various kinds of connections, draws on the work of Robert Jay Lifton in *The Broken Connection*, although I depart from his categories in a number of ways. See Lifton (1983).

[17] I add "familial" to Lifton's "biological," for we should include adopted children and other forms of non-biological family relationships. One could argue, however, that non-biological relations would belong in the third category, "human relatedness," which would allow this to be a separate category for biological connections, which are deeply important for Confucians. If we extend this category, we could include larger structures that provide a sense of membership and belonging, such as tribes, ethnic groups, or even nations. The Confucians themselves see the ideal society as a family writ large, and use familial metaphors to describe the proper relationship between rulers and subjects.

[18] Throughout East Asia, the ancestral altar within the home serves as a visible reminder of our place in the extended family line. Chinese families will often consult the ancestors on important matters and inform them of significant passages or decisions. The practice of "feeding" the ancestors and giving them money and other useful objects (through ritual burning) symbolically conveys the sense that our ancestors' "survival" depends on our remembering.

2. *Creative*: This can be seen in what we individually produce and also in our participation in and contribution to an ongoing creative community or tradition. We are engaged in larger projects that transcend and survive us. Confucians, but not Zhuangzi, emphasize this dimension. Confucians point out that it *matters* to us what happens to our work, values, and ideas (which, at some level, we feel to be an extension or dimension of our "self") after our deaths.[19]
3. *Human-relatedness*: This is the reciprocal influence people have on each other through their relationships, and looks at the effects we have on the lives of people like students, family, and friends. These themes are richly developed in the Confucian tradition.
4. *Nature itself*: This category looks at our participation in the natural world. While this is largely a Daoist theme, it is also found in Confucian thought. In Zhuangzi's deathbed experience, it is this form of connectedness that provides solace. Human beings can take comfort in the fact that our lives and deaths are intimately bound up in the cyclical processes of nature.
5. *Experience of oneness or nonduality*: The realm of the "mystical" is seen most clearly in Zhuangzi. The actual experience is often equated with a particular psychic state, what might be called an "expanded state of consciousness." Such a state is described in almost every religious tradition, and normally has the qualities of nonduality, where the boundaries between self and world, subject and object, disappear.

The modes of connectedness and continuity emphasized by Confucians are family, work, human-relatedness, and tradition/history. They are sustained by learning, ritual, and text. *They are grounded in the narrative conception of temporality and selfhood, and depend on our ability to remember.* By placing ourselves in the context of the larger familial and social connections that constitute and sustain us, we understand how our life and death fit into larger systems of meaning, systems that existed before our birth and that will continue after our death. We understand that our "name"—our character and accomplishments—will persist after our death and will continue to educate and inspire others.

In contrast with the Confucian picture, *the connections at the heart of Zhuangzi's vision are grounded in momentary temporality and no-self, and depend on our ability to forget.* Whereas Confucian connections are made between highly cultivated selves, Zhuangzian connections are made when we "match up nature with nature." The more we learn to forget our culturally constructed selves—and forget tradition, norms, roles, and language—the freer and easier our lives will be. A powerful awareness of transience and change helps us to avoid the errors of ossification and holding on.

[19] All of the Confucian thinkers emphasize that while the body does not remain, the name (*ming*) does. Confucius said, "The gentleman hates not leaving behind a name when he is gone" (15.20). This is in stark contrast with Zhuangzi's statement that "the sage has no name." In addition, the tradition of which the deceased was a part, and which he has done his best to preserve and transmit, remains. Confucians saw themselves as the keepers of the humanizing tradition. Mencius said, "All a gentleman can do in starting an enterprise is to leave behind a tradition which can be carried on" (1B14).

One way to understand Confucian and Zhuangzian conceptions of self and no-self is through the lens of "remembering/forgetting" or "holding on/letting go." Within the Chinese context, it becomes clear that achievement of selfhood depends on remembering, while no-self is achieved through forgetting. Both perspectives lead to ways of coming to terms with death.

There are multiple senses in which memory plays an important role in Confucian thought. Memory is essential for the individual to construct a "self" over time (keeping in mind that this individual life narrative is embedded within the family narrative, which is why Confucians are connoisseurs of genealogy). Beyond this, the cultural memory, preserved through texts and ritual, is our collective endowment that allows us to realize our humanity.[20]

Zhuangzi, on the other hand, offers a way of forgetting. This can be achieved through meditative practices Zhuangzi calls *zuo wang,* "sitting in forgetfulness," and *xin zhai*, "fasting of the mind," and also through absorption in skillful activity.[21] Zhuangzi advises, "Forget things, forget heaven, and be called a forgetter of self. The man who has forgotten self may be said to have entered heaven" (W, p. 133). "Fish forget themselves in the rivers and lakes; people forget themselves in the arts of the Way" (p. 61; see also pp. 48–49).

Conscious remembering—through ritual, tradition, and text—is essential for creating the Confucian good life but is an impediment to living well for Zhuangzi. As we have seen, Confucian remembering and Zhuangzian forgetting both provide means by which human beings can experience profound, life-enhancing connections.

2.6 Facing Our Deaths: Contemporary Implications

2.6.1 The Differing Perspectives of Confucian and Daoist Thought

Although a great historical and cultural distance separates us from Confucius, Xunzi, and Zhuangzi, their insights and perspectives can speak to us today. One can imagine their commentary on and contributions to certain developments in the contemporary American treatment of the dying and of death.

[20] The child in a Confucian home was educated by memorizing the classics. The importance of history for Confucians demonstrates the role that cultural memory plays. As Confucius explicitly acknowledges, preservation and transmission are more important than innovation. Since, for Confucians, the ideal way resides in a past Golden Age, preserving the cultural memory is a sacred obligation. Furthermore, ritual is a way of inscribing memory on the body. The body is humanized through, crudely put, a type of "muscle memory" so that one moves, feels, and responds according to the Way that has been passed down. In other words, the expression of memory is itself a bodily phenomenon.

[21] See Ivanhoe (1993).

There are three developments in the American approach to death that can be illuminated by the insights of the Chinese thinkers. From the Confucian perspective, we can reflect on the importance of greater family involvement in the treatment of loved ones through the dying and post-death processes. From the Zhuangzian perspective, we can look at the movement toward more natural burials and the attempt to cultivate greater equanimity in the face of imminent death through fostering a dissolution of boundaries between self and world.

2.6.2 *Family Involvement*

The Chinese thinkers had to address the issue of death because it surrounded them constantly. Death and dying are rarely actually seen in contemporary American society (other than on a screen). The stark reality of the public charnel ground or funeral pyre, or even the more intimate experience of an at-home death, are simply not significant parts of the American landscape. Death usually takes place in removed, institutional settings, walled off from our everyday life. Robert Blauner observes:

> Hospitals are organized to hide the facts of dying and death from patients as well as from visitors. Sudnow quotes a major text in hospital administration: "The hospital morgue is best located on the ground floor and placed in an area inaccessible to the general public. It is important that the unit have a suitable exit leading on to a private loading platform which is concealed from hospital patients and the public." (Blauner 1976, p. 43)[22]

A common view in the medical profession is that death is a failure. The laws against euthanasia in most states have, as their implicit foundation, the belief that it is always better, under any circumstances, to prolong life. Doctors succeed when they keep a patient alive, fail when the patient dies. On this way of thinking, there is no notion of dying well, of a "good death." Yet, in the works of both Zhuangzi and the Confucians, as well as other religious traditions of East and South Asia, there are numerous examples of masters who have died in exemplary ways. One of the first lessons we can learn, then, is that it is possible to die well, even beautifully. But, as Ira Byock points out, "Good deaths were not random events or matters of luck; they could be understood and, perhaps, fostered" (Byock 1997, p. 31).

---

[22] The remarkable success of medical science in extending life has also enabled this denial of death. Patricia Anderson observes that until the nineteenth century, one-half of the people on the planet died before reaching the age of 8: "Everyone saw death happen frequently, and both the event and our sense of it were integrated into daily life.... Then suddenly, about 50 years ago in the West, a dramatic transformation occurred.... The average life span soared and we stopped seeing people die. For the first time in history, a human child could grow to full adulthood without experiencing the death of anyone close to him or her.... Death became a possibly curable disorder" (Anderson 1996, p. 5).

Before the twentieth century, most people died at home. Throughout the twentieth century and to the present day, the dying have usually been moved to hospitals or other institutions, perhaps as much from our own need to hide from death as from our sense that it is the best thing for the dying person. The modern American way of death, incisively criticized by Jessica Mitford, has been characterized by professionalization and medicalization. For many Americans, death has become the domain of specialists, and families are no longer as directly involved in preparing the bodies of their loved ones, burying them, or organizing an opportunity for collective mourning.[23]

Due to the efforts of the hospice movement, this is changing. The position of this movement, reflected in Confucius' own deathbed story, is that one should die surrounded not by those in official roles, but by loved ones. Increasingly, friends and family, rather than doctors and nurses, are playing a significant role in the dying process. While the participation of health care professionals is essential, a Confucian perspective would encourage us to again make dying the responsibility of loved ones. Hospice nurse Chris Sonnemann makes the point that the act of caring for loved ones during and after their deaths provides substantial benefits to both the dying and the survivors. The act of handling the dead body (e.g., cleansing, preparing), which is done by family members in many cultures but handled by professionals in America, can help people come to terms with grief. She writes: "People who have spent weeks, months or sometimes years caring for an ill person have done much for their loved one. Caring for them after death is a natural progression in completing the circle of care.... I have noticed that families who care for their own dead have an easier time with their grief" (Sonnemann, in Carlson 1997, p. 36).

An increasing number of Americans are rejecting the modern professionalized model of care for the dying and dead in favor of what Kathleen Garces-Foley and Justin Holcomb call the "post-modern funeral." In an effort to re-personalize death, people are seeking to participate more in the funeral process and the memorialization of their loved ones. This new approach rejects the complete professionalization of death rituals, and gives loved ones a more active role. More Americans have returned to the practice of participating in the preparation of the body through washing and dressing it. People are also having home viewings and home funerals, a natural extension of the wish of more people to die at home. Garces-Foley and Holcomb write, "Postmodern funerals have a strong communal function not only in the planning stages, but also in their execution. These services invite participation from friends and family and through shared eulogizing create a sense of community among the bereaved" (Garces-Foley and Holcomb 2006, p. 227).

---

[23] Beginning in the 1950s, most Americans died in some kind of institutional setting rather than at home. The percentage dying in institutions rose throughout the twentieth century. Recent studies have shown that although 80% of Americans would prefer to die at home, only 20% actually do (Stanford University School of Medicine).

2.6.3 Back to Nature

There is also increasing interest in more natural, environmentally conscious, and meaningful ways of disposing of the corpse. Some people who favor burial are avoiding embalming and the use of elaborate caskets, and choosing so-called "green" or "woodland" burials instead, a choice that Zhuangzi would certainly affirm. These often involve burying someone in the woods, often where a tree can be planted over the grave. People are also buried in easily decaying material, such as a shroud or a biodegradable casket. People who opt for this kind of funeral see it as a way to most directly return the corpse to the natural cycle.

Some consider the next step in the evolution of burial to be "human composting." Katrina Spade received an environmental fellowship for her work designing a facility for human composting, part of what she calls her "Urban Death" project. Spade points out that there is no reason why a nutrient-rich human body cannot serve as compost out of which new life can grow; composting is already done with livestock bodies on some farms. The dead body is placed inside "carbon-rich material, like wood chips or sawdust," and the addition of extra moisture and nitrogen leads to microbial activity that begins breaking down tissue and creating a "soil-like substance." When the compost is ready, the loved ones would collect some of it to use in their gardens, with the rest going to parks or conservation lands. The cost is far less expensive than a conventional American burial. Echoing Zhuangzi, Spade writes that this approach is a way of "connecting death to the cycle of nature," which will "help people face their own mortality and bring comfort to the bereaved" (Spade, in Einhorn 2015, p. D1).[24]

2.6.4 Loss of Self and Sense of Oneness

Zhuangzi's work is permeated by a sense that all boundaries are constructed, and he uses a number of techniques to deconstruct that conception of self that most people—and certainly Confucians—hold. As we have seen, Zhuangzi suggests that those who have "lost" or "forgotten" their self can face death—even imminent, gruesome death—with equanimity, humor, and joyful acceptance.

If we believe there is truth to this, then it might be possible to overcome death anxiety through practices (e.g. contemplative, meditative) that help us let go of the self that we normally spend so much time building up and defending. In his study of mystical experience, John Clarke argues that such experiences generally involve the following characteristics: "an experience of oneness, ... an attenuation of the sense

[24] A bill currently before the Washington State Legislature would legalize human composting, making it the first state in the nation to do so (Johnson 2019).

of dualism of subject and object, the lessening or loss of one's sense of one's separate existence, a feeling of bliss and harmony" Such an experience, he states, "nullifies all fear of pain and death" (Clarke 1994, p. 374).

However, it is difficult to achieve this type of consciousness without cultivation. While those who have prepared (e.g., meditated for years) may be able to face death with equanimity, what about those whose lives have not involved such practices or experiences, and who find themselves facing death with terror and uncertainty? For most of us, it would be too difficult to "forget" and let go after a lifetime of holding on. Very little work is being done in the area of how to help people die well (other than to provide for their physical comfort, which is still far too rare). Manish Agerwal observes, "We don't die well in this country. And we have pretty limited tools to help people deal with their fear. Prozac doesn't work. The issue isn't depression; it's facing your mortality" (Pollan 2018).

One promising avenue, the use of consciousness-expanding substances ("psychedelics" or "entheogens") for terminally ill patients, was explored in the 1960s and 1970s, but virtually all research and experimentation in this area was closed off for political reasons. An increasing number of studies, some published in peer-reviewed science journals, suggest that mitigating death anxiety in terminally ill patients might be possible through the use of psychedelics with a guide in a carefully controlled environment.

Bestselling author Michael Pollan, who recently wrote a book on the therapeutic potential of psychedelics and himself took them under the care of a therapist/guide, wrote the following statement, which echoes the perspective of Zhuangzi:

> That I could survive the dissolution of my ego and its defenses was surely something to be grateful for.... For the first time, I began to understand what the volunteers in the cancer-anxiety trials were telling me—how it was that a single psychedelic journey granted them a perspective from which the very worst life can throw at us could be regarded objectively and accepted with equanimity. (Pollan 2018)

A study conducted by Walter Pahnke at Sinai Hospital in Baltimore, reached the following conclusions about the changes in many of the patients who took psychedelics:

> Most striking was the decrease in the fear of death. It seems as if the mystical experience, by opening the patient to usually untapped ranges of human consciousness, can provide a sense of security that transcends even death. (Pahnke 1969, p. 12)

> In fact, the experience of deeply felt positive mood may be more the result than the cause of this change in attitude toward death. Our data show that these feelings are released most fully when there is complete surrender to the ego-loss experience. (Pahnke 1969, p. 15)

For many patients, experiences of ego-dissolution brought about greater senses of oneness and connection, and produced numerous positive changes, including a decrease in death anxiety. A study done at Johns Hopkins concluded that the use of psilocybin, when combined with meditation practice, can have powerful, lasting psychological effects: "Participants showed significant positive changes on longitudinal measures of interpersonal closeness, gratitude, life meaning/purpose, forgiveness, *death transcendence*, daily spiritual experiences, religious faith and coping" (Griffiths et al. 2018, p. 49; italics mine).[25]

2.7 Conclusion

The great Sinologist David Keightley wrote, "Nowhere (in the Chinese texts) do we find the epic concerns of the Iliad and Odyssey, which focus on the manner of death, the ritual treatment of the dead, and the unhappy fate of the shades after death and which express so powerfully the tragic ... poignancy that death confers on the human condition.... The very silence of the Chinese texts about such matters suggests a remarkable Chinese ability to emphasize life over death" (Keightley 1990, p. 33).

From what we have seen, it is clear that the Chinese texts are certainly not silent on issues related to death and dying, even in the areas Keightley mentions. However, Keightley is directing our attention to the revealing ways that the Chinese attitudes toward death differ from those of ancient Greece.[26] Keightley rightly points to the lack of much Chinese speculation about death in the texts: "Nowhere do we find a philosophical discourse like Plato's *Phaedo*, that is devoted to the nature of death and soul" (Keightley 1990, p. 33). There is not much discussion of the "nature of death," the "metaphysics of death," or the "nature of the afterlife." There is no thanatology in early Chinese thought.

The Chinese texts focus on the ways that death actually enters our lives: our thoughts about death, our experiences of the death of others, and our own experience of dying. So in this sense, the Chinese are both this-worldly and very interested in death and dying. Death permeates our lives in its guise as transience, finitude, and loss.

The Confucians and Zhuangzi face death and loss in radically different ways. In his poem "Ash Wednesday," T.S. Eliot writes, "Teach us to care and not to care" (Eliot 1971, p. 60). This is another way to approach the divide between the Confucians and Zhuangzi. The Confucians teach us to care, to take things seriously. Zhuangzi teaches us not to care, to take things lightly. Eliot's words convey the importance of incorporating both perspectives into our lives.

---

[25] One participant in a New York University study of death anxiety in cancer patients experienced "debilitating end-of-life anxiety from the moment he was diagnosed until the day he ingested psilocybin, extracted from hallucinogenic mushrooms while laying on a psychiatrist's couch." Gabrielle Agin-Liebes, the research manager for the NYU study, told *The Atlantic* that this patient "had one of the highest ratings on the anxiety scale that we had seen: 21 out of 30. To qualify for the study you only need an eight. The day after his first dosing session, he dropped to zero, and for 7 months he's stayed there. Zero anxiety" (Agin-Liebes, in Morin 2014).

[26] Some of these arise from the difference in context in which the treatment of death is embedded in the texts. Given the Greek emphasis on the military hero (in the Iliad) and the wandering adventurer (in the Odyssey), there is a corresponding emphasis on virtues that are martial (including skill in battle and military leadership) and involve levels of craftiness and survival skills. Such an approach is not found as much in the Chinese texts, in which the mythological heroes are praised more for their skill with the brush or hoe than the sword and shield. Early Chinese mythology features cultural heroes far more than military heroes.

Throughout the last two millennia, Confucians and Daoists have learned from each other, argued, vied for influence, and given rise to syncretic expressions. But neither ever "triumphed" over the other, nor did most of them try. Many Chinese have maintained a connection with both traditions throughout their lives, in part because both give insightful perspectives on human existence and present compelling visions of how to live. We can also draw on the insights of both traditions as we think about death.

Each perspective points us to important dimensions of existence and expresses powerful truths. The Confucians set our life and death against a backdrop of larger social realities—family, politics, history. The Daoists set our life and death against a backdrop of the larger natural order, the unfolding cycles of nature and rhythms of the Dao. The Confucians show us what it means to be a *social* being; Zhuangzi reminds us that we are essentially *natural* beings. Confucians show us how important it is to cultivate the self, and Zhuangzi shows us how important it is to forget the self.

Confucians show us that it is important to step back and see the course of our development over time, our coming into selfhood. This requires remembering. At the same time, Zhuangzi reminds us, we reach the greatest heights when self and time disappear. This is a form of forgetting. We must live in partial remembering and partial forgetting, for each must be tempered by the other. We must both cultivate ourselves over the course of our lives and forget ourselves in the greatest moments of our lives. Living well, then, requires that we live the continuous tension of remembering and forgetting, holding on and letting go, cultivating the self and forgetting the self.

By seeing how these important Chinese thinkers understood, approached and learned from death, we understand better what they thought about life, and why they lived as they did. Hopefully, their words can reach across a great cultural and historical distance to help us live better lives—and die better deaths.

References

Anderson, Patricia. 1996. *All of us: Americans talk about the meaning of death*. New York: Dell Publishing.

Bellah, Robert. 1970. Christianity and symbolic realism. *Journal for the Scientific Study of Religion* 9 (2): 89–96.

Berkson, Mark. 2011. Death in the *Zhuangzi*: Mind, nature and the art of forgetting. In *Mortality in traditional Chinese thought*, ed. P.J. Ivanhoe and Amy Olberding, 191–224. Albany: SUNY Press.

———. 2016. Xunzi as a theorist and defender of ritual. In *The dao companion to the philosophy of Xunzi*, ed. Eric Hutton, 229–268. New York: Springer. 2016.

Blauner, Robert. 1976. Death and social structure. In *Death and identity*, ed. Robert Fulton, 35–58. Bowie: Charles Press Publishers.

Byock, Ira. 1997. *Dying well: Peace and possibilities at the end of life*. New York: Riverhead Books.

Carlson, Lisa. 1997. *Caring for the dead: Your final act of love*. Hinesburg: Upper Access Book Publishing.

Clarke, John. 1994. Mysticism and the paradox of survival. In *Language, metaphysics and death*, ed. John Donnelly, 367–382. New York: Fordham University Press.

Confucius. 1979. *The analects*. Trans. D.C. Lau. New York: Penguin.

Einhorn, Catrin. 2015. Returning the dead to nature. *New York Times*, April 13.

Eliot, T.S. 1971. *Collected poems and plays*. Orlando: Harcourt Brace Jovanovich.

Freud, Sigmund. 1989. *An outline of psychoanalysis*. Trans. J. Strachey. New York: W.W. Norton.

Garces-Foley, Kathleen, and Justin Holcomb. 2006. Contemporary American funerals: Personalizing tradition. In *Death and religion in a changing world*, ed. Kathleen Garces-Foley, 207–227. Armonk: M.E. Sharpe.

Gordon, Rosemary. 1978. *Dying and creating: A search for meaning*. London: Society of Analytical Psychology.

Graham, Angus C. 1989. *Disputers of the dao: Philosophical argument in ancient China*. LaSalle: Open Court.

Griffiths, Roland, Matthew Johnson, William Richards, et al. 2018. Psilocybin-occasioned mystical-type experience in combination with meditation and other spiritual practices produces enduring positive changes in psychological functioning and in trait measures of prosocial attitudes and behaviors. *Journal of Psychopharmacology* 32 (1): 49–69.

Ivanhoe, Philip J. 1993. Skepticism, skill and the ineffable tao. *Journal of the American Academy of Religion* 61 (4): 639–654.

———. 2003. Death and dying in the analects. In *Confucian spirituality*, ed. Tu Wei-Ming and Mary Evelyn Tucker, 222–233. New York: Crossroads.

Johnson, Kirk. 2019. Washington state weighs new option after death: Human composting. *New York Times*, Jan 26.

Keightley, David N. 1990. Early civilization in China: Reflections on how it became Chinese. In *Heritage of China*, ed. Paul S. Ropp, 15–54. Berkeley: University of California Press.

Kjellberg, Paul, and P.J. Ivanhoe, eds. 1996. *Essays on skepticism, relativism and ethics in the Zhuangzi*. New York: SUNY Press.

Lifton, Robert. 1983. *The broken connection: On death and the continuity of life*. New York: Basic Books.

Mencius. 1970. *Mencius*. Trans. D.C. Lau. New York: Penguin Books.

Morin, Roc. 2014. Prescribing mushrooms for anxiety. *The Atlantic*, April 22. Https://www.theatlantic.com/health/archive/2014/04/chemo-for-the-spirit-lsd-helps-cancer-patients-cope-with-death/360625/.

Pahnke, Walter N. 1969. The psychedelic mystical experience in the human encounter with death. *The Harvard Theological Review* 62 (1): 1–21.

Paper, Jordan, and Lawrence Thompson, eds. 1998. *The Chinese way in religion*. New York: Wadsworth Publishing.

Pollan, Michael. 2018. My adventures with the trip doctors. *New York Times Magazine*, May 15. Https://www.nytimes.com/interactive/2018/05/15/magazine/health-issue-my-adventures-with-hallucinogenic-drugs-medicine.html.

Potter, Peter. 1988. *All about death*. New Canaan: William Mulvey.

Rappaport, Roy. 1999. *Ritual and religion in the making of humanity*. Cambridge, UK: Cambridge University Press.

Rosenblatt, Paul, Patricia Walsh, and Douglas Jackson. 1976. *Grief and mourning in cross-cultural perspective*. New Haven: HRAF Press.

Slingerland, Edward. 2007. *Effortless action: Wu wei as conceptual metaphor and spiritual ideal in early China*. New York: Oxford University Press.

Stanford University School of Medicine. n.d.. Where do Americans die? https://palliative.stanford.edu/home-hospice-home-care-of-the-dying-patient/where-do-americans-die/. Accessed 20 Feb 2019.

Xunzi. 1963. *Hsün Tzu: Basic writings*. Trans. B. Watson. New York: Columbia University Press.

Yearley, Lee. 1983. The perfected person in the radical Zhuangzi. In *Experimental studies on Chuang-Tzu*, ed. Victor Mair. Honolulu: University of Hawaii Press.

———. 1995. *Facing our frailty: Comparative religious ethics and the Confucian death rituals. The Gross Memorial Lecture for 1995*. Valparaiso: Valparaiso University Press.

Zhuangzi. 1964. *Chuang Tzu: Basic writings*. Trans. B. Watson. New York: Columbia University Press.

———. 1989. *Chuang Tzu: The inner chapters*. Trans. A.C. Graham. Boston: Unwin Paperbacks.

———. 1994. *Wandering on the way: Early taoist tales and parables of Chuang Tzu*. Trans. V. Mair. New York: Bantam Books.

Chapter 3
Secular Death

Amy Hollywood

Abstract This essay explores how those in the modern West who live outside traditional religious communities might approach death—our own and those of others. It suggests that difficult literature—reading it and writing it—might be a kind of training ground for living with, and in the face of the difficulty of, death itself.

> Starting from nothing with nothing when everything else has been said (Howe 2010)

This single, unpunctuated line is the second paragraph of Susan Howe's *That This*, a book-length poem in four parts dedicated to the memory of her third husband, the philosopher, Peter H. Hare. She has just described her discovery of his dead body: "I knew when I saw him with the CPAP mask over his mouth and nose and heard the whooshing sound of air blowing air that he wasn't asleep. No" (Howe 2010, p. 11). Howe juxtaposes her refusal—"No"—with Sarah Pierpont Edwards' letter of July 3, 1758 announcing the death of her husband, Jonathan Edwards, to one of their daughters.

> Oh that we may kiss the rod and lay our hands on our mouths! The Lord has done it. He has made me adore his goodness, that we had him so long. But my God lives; and he has my heart ... We are all given to God: and there I am, and love to be. (Howe 2010, p. 12–13)

"I admire," Howe writes, "the way thought contradicts feeling in Sarah's furiously calm letter" (Howe 2010, p. 13).

The furiously calm contradictions of Sarah Edwards' letter find their justifying source in the anguished Psalm to which her words likely allude. Psalm 2 opens with a question: "Why do the heathen rage, and the people imagine a vain thing?" The answer, in the King James version, defies summation. Reading the Psalm, I am not

---

This essay was first published under the title "Secular Death" in *Harvard Divinity Bulletin*, Summer/Autumn 2016, pp. 20–30.

A. Hollywood (✉)
Harvard Divinity School, Cambridge, MA, USA
e-mail: ahollywood@hds.harvard.edu

T. D. Knepper et al. (eds.), *Death and Dying*, Comparative Philosophy of Religion 2, https://doi.org/10.1007/978-3-030-19300-3_3

always sure who is speaking to whom, but there is no doubt as to what is being promised:

> Ask of me, and I shall give thee the heathen for thine inheritance, and the uttermost parts of the earth for thy possession.
> Thou shalt break them with a rod of iron; thou shalt dash them in pieces like a potter's vessel.
> Be wise now therefore, O ye kings: be instructed, ye judges of the earth.
> Serve the LORD with fear, and rejoice with trembling.
> Kiss the Son, lest he be angry, and ye perish from the way, when his wrath is kindled but a little. Blessed are all they that put their trust in him.

This is difficult writing: the rage and the violence seem directed—everywhere. Sarah Edwards feels this rage and this violence; she feels it directed against her husband and against herself and against her daughter, but she also, with ease, identifies the rod, which will "dash them to pieces like a potter's vessel," with the Son and so with the promise of blessedness.

For Sarah Edwards, the one who destroys is also the one who saves. Howe sees this clearly:

> For Sarah all works of God are a kind of language or voice to instruct us in things pertaining to calling and confusion. I love to read her husband's analogies, metaphors, and similes.
>
> For Jonathan and Sarah all rivers run into the sea yet the sea is not full, so in general there is always progress as in the revolution of a wheel and each soul comes upon the call of God in his word. (Howe 2010, p. 12)

Scripture calls and gives a calling to Jonathan and Sarah Edwards. It calls them to God where, Sarah writes, "I love to be." Howe sees this in Jonathan and Sarah Edwards, but she doesn't experience it: "I read words but don't hear God in them" (Howe 2010, p. 12). Howe loves Jonathan Edwards's analogies, metaphors, and similes, most of them biblically based, but she hears in these words "the unpresentable violence of a double negative." The rod, perhaps, without the Son? Or more troubling still, the rod with the Son but without any promised salvation? For Howe, what is crucial is how unpresentable this is.

That This goes on to perform that unpresentable violence in often illegible textual collages, strips of words, and fragments of words cut from the writings of the

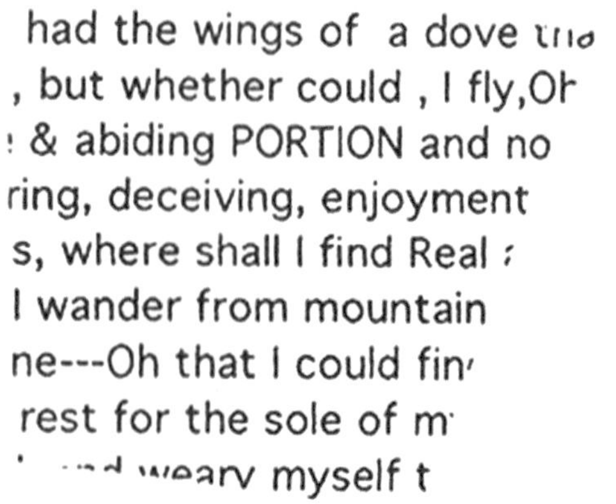

Fig. 3.1 Page 47 of Susan Howe's *That This* (By Susan Howe, from THAT THIS, copyright ©2010 by Susan Howe. Reprinted by permission of New Directions Publishing Corp.)

had the wings of a dove
, but whether could , I fly,Oh
& abiding PORTION and no
ring, deceiving, enjoyment
s, where shall I find Real
I wander from mountain
ne---Oh that I could fin
rest for the sole of m
weary myself t

Edwards family and other, unspecified texts, placed seemingly at random on the page. The Psalms appear again, in a more hopeful light, although one that remains tethered to Howe's framing question: "where shall I find Real" (Fig. 3.1).

"Maybe," Howe writes, "there is some not yet understood return to people we have loved and lost. I need to imagine the possibility even if I don't believe it" (Howe 2010, p. 17). These poems mark the impossible necessity of imagination; Howe returns to her husband—she returns to the Edwards family—through these barely legible textual collages. Her reading and her writing—of her husband's autopsy report, his emails, the detritus that catches her eye as she wrestles with the reality of his death, the paintings of Poussin and the art historian T. J. Clark's reflections on those painting—and what she makes of them *are* the return:

> This sixth sense of another reality even in simplest objects is what poets set out to show but cannot once and for all.
> If there is an afterlife, then we still might: if not, not. (Howe 2010, p. 34)

This is the background against which I want to talk about how we die now—those of us who are no longer Christian. I am interested in how we approach the difficulty of death, our own and others, and how difficult literature, writing it and reading it, might be a training ground for approaching the difficulties of death—and of life. I will speak in terms of what Michael Warner calls "ethical secularism," a version of secularism that asks how we live now, those of us for whom religious practices are no longer formative. In Warner's words, "it presents itself as a project for becoming the kind of person who can rightly recognize the conditions of existence, and although it is an attempt to overcome Christianity it does not secure its stance as a privileged default against the particularities of religion." I'd like to say that I don't wish to overcome Christianity. I'd like to say that death demands particularity. But for me, as for many others, Christianity is simply—but there is nothing simple about it—gone.

When I think about religion and what we grandiously call literature, I am less interested in the exploration of religious themes or images—although they are strewn through the work of those I'll discuss here—than in the analogies and disanalogies between literary practices and religious practices of writing and reading. The practices of religious elites—and increasingly not only elites—in Western Christianity circled endlessly around the Psalms, with reading, recitation, song, and meditation leading to the production of new songs in and through engagement with the biblical text. In medieval monastic life, monks and nuns, either individually or in community, enact the continual praise of God that is heaven itself through their recitation or singing of the Psalms. For John Cassian, a key figure in the development and theorization of Christian monasticism in the Latin West, the techniques of repetition central to the monastic life are not at odds with, but themselves produce, spontaneity.

Dying and dead monks and nuns are surrounded by their communities, who recite or sing the Psalter over them. By the high and late Middle Ages, these practices have moved out of the confines of the monastery. The Beguines, semi-religious women living lives dedicated to God and Christian perfection in the world, were often employed to care for the sick and the dying. This included reciting or singing the Psalms over them, perhaps at times with them. (Did they sing in Latin? The

French, Flemish, German, Italian, English vernaculars? Many of the dying may not have understood the words of the song.)

Those raised within Christian and Jewish traditions often forget how *difficult* the Psalms are—not just formally, but also in their vivid and incredibly complex imagery, in the harshness of the world and the God they depict, in the truths they purport to tell. For early theologians, theorists of the Psalms and literary critics all, and often poets, the Psalms are at the heart of the Christian life because they contain the entire range of human emotion. Through uttering the words of the Psalms and looking from the book of scripture to what Bernard of Clairvaux calls the book of experience, moving back and forth between the two so that the words of the Psalmist become one's own and emerge, spontaneously, from one's lips, Christians come to understand who they are, who God is, how they stand, individually and communally, before, with, and in that God.

The Psalms are not my scripture. But since I was a small child I have *read* religiously, perhaps I have even begun to write that way. And that is what I see, vividly, complexly, difficultly, in the work I love, for example this (a)typical juxtaposition, from Fanny Howe's (Susan Howe's sister—or is Susan the sister of Fanny?) *Tis of Thee* (2003), chosen (almost) at random.[1]

> X: and Z:
> Any discussion of race is really a discussion about the creation of the universe.
> [page break]
> Y:
> Now I believe that when the Messiah comes the world will have no images,
> since the image will be cut free
> from the object, released like beef from a cow,
> and competition will automatically founder
> as an instinct, having no visible object in sight.
> Then on that day I won't have to look for you in order to know you. (Howe 2003, pp. 60–61)

I don't believe in the messiah, but I believe in Fanny and Susan Howe. What does that make me?

***

I have a fantasy, that when I am dying, someone will read to me, the opening pages of *The Portrait of a Lady*.

> Under certain circumstances there are few hours in life more agreeable than the hour dedicated to the ceremony known as afternoon tea. There are circumstances in which, whether you partake of the tea or not—some people of course never do—the situation is itself delightful. Those that I have in mind in beginning to unfold this simple history offered an admirable setting to an innocent pastime. The implements of the little feast had been disposed upon the lawn of an old English country house in what I should call the perfect middle of a splendid summer afternoon. Part of the afternoon had waned, but much of it was left, and what was left was of the finest and rarest quality. Real dusk would not arrive for some hours; but the flood of summer light had begun to ebb, the air had grown mellow, the shadows were long upon the smooth, dense turf. They lengthened slowly, however, and the

[1] X = African American man; Y = European American woman; Z = their grown son.

> scene expressed that sense of leisure still to come which is perhaps the chief source of one's enjoyment of such a scene at such an hour. From five o'clock to eight is on certain occasions a little eternity; but on such an occasion as this the interval could be only an eternity of pleasure. (James 1975, p. 17)

There are three men on the lawn in Henry James's not so very innocent scene, an old man and two considerably younger—a father, a son, the son's friend. (And two dogs, a collie and a "bristling, bustling terrior" who only later, in passing, will we find to be named Bunchy. All of the names come later.)

Daniel Touchett and his son Ralph are both dying.[2]

Their history, and that of the young woman who comes to join them, may seem simple, even banal. How can one speak of eternity with respect to an afternoon tea? (A little eternity and hence no eternity at all; and yet, beyond the irony and the boredom, on *this* occasion "an eternity of pleasure.")

I found myself thinking about this scene in James and my own imaginary deathbed scene—during which I am remembering James or listening to some unspecified person reading James to me—while reading, entirely by accident, David Orr's *Beautiful & Pointless: A Guide to Modern Poetry*. Orr wants to argue against the notion that poetry needs to be something grand, dealing with the sublime and the eternal, in order to be interesting and worth spending one's time on. Poetry, he writes "seems beautifully pointless, or pointlessly beautiful, depending on your level of optimism" (Orr 2012, p. 187). (An afternoon tea on a beautiful lawn dipping down to the Thames. James refuses to leave its beauty alone.)

Orr's line marks the end of a section; a new one begins, the penultimate one of the book.

> My father died of cancer in March of 2007, as I was beginning work on this book. He was sixty-one. It's difficult to type those sentences for many reasons, not least among them the fact that I've been a book critic for over a decade now, and almost always find myself cringing during the inevitable fetch-me-a-tissue moment in any personal essay or memoir. Still, throughout this book … I've tried to suggest what a relationship with poetry actually looks like, in both its limitations and strengths. I've described it as a private pleasure and an occasional irritation that can't easily be justified in public terms. Having said this, I'd be falling short if I didn't try to offer some sense of what—for me—poetry has proven it can and cannot give. Sad as it may be, we often discover the true contours of any relationship in the situations that matter most to us; and sadder still, those situations tend to be ones in which something we love is lost, or in danger of being lost. So pull out your tissues, and let's talk about it. (Orr 2012, pp. 187–188)

(I've got a lot of these stories. My father, dead at 65, when I was 23. My oldest brother and mother dead 9 years later, in Pedro's case almost to the day of my father's death. My mother 4 days later. My brother Michael died in 2013 on the same date as Pedro. Between them, a sister and another brother, my best brother, Daniel. And that's not all.)

(I couldn't give a shit about tissues.)

---

[2] Editors' note: Daniel Touchett and his son, Ralph, are the father and son, whom the author refers to above, in Henry James' *The Portrait of a Lady*.

"Cancer," Orr writes, "can kill you in many ways" (Orr 2012, p. 188). In the midst of and likely caused by chemotherapy, his father had a stroke that left him partially paralyzed and unable to shift the pace, intonation, or stress of his words. Suffering from what is sometimes called "flat affect" (surely a misnomer—the voice is flat, but is the affect? perhaps it's just easier to think that what we can't hear isn't there), Orr's father's couldn't "slow down and modulate his voice." A speech therapist gave him various exercises meant to help.

(Cancer can suffocate you; rot you from the inside out or from the outside in; make you hurt so badly you can't even scream. It can give you a heart attack or tangle your guts so that you heave your own shit. No one wants to know all the ways that cancer can kill you.)

As Orr self-deprecatingly explains, the not-unself-interested thought occurred to him that poetry might also help his father. It deals, often, with—or perhaps better in—the stress of meter, the intonations engendered by rhythm and rhyme, the pacing needed to articulate assonances and alliterations. He searched his parents' house for leftover books from college, and "flush with inspiration" returned to his father "armed with the fruits of English poesy." He tells us he learned one lesson very quickly:

> Do not attempt to get a stroke victim to read Hopkins. "I caught this morning morning's minion, king-/dom of daylight's dauphin, dapple-dawn-drawn Falcon …" I can barely pronounce this myself, and I have full use of my tongue. We did a little better with Robert Frost. (Orr 2012, p. 190)

(I gave my father *The Complete Poems of Gerard Manley Hopkins* for Christmas a year or two before he died. He'd had an old, mottled edition, its wartime paper disintegrating under my hand as I tried to read it. Whose gift was this?)

Orr and his father read from Frost's "The Silken Tent"; Orr cites the opening four lines of the poem.

> She is as in a field a silken tent
> At midday when the sunny summer breeze
> Has dried the dew and all its ropes relent,
> So that in guys it gently sways at ease …

Orr reflects on what he loves in the poem; its technical virtuosity—the entire poem is one sentence—and what he calls "the delicate exactness of the first line." "'She *is as in* a field a silken tent,' rather than, for instance, 'She *is like* a silken tent in a field.'" For Orr, following the critic Robert Pack "'the metaphor of the tent does not merely describe the 'she' of the poem, but rather the relationship between the speaker and the woman observed'" (Orr 2012, p. 191). (I am not at all sure that this is right.) What Orr likes about it is that Frost, in giving a relation—the metaphor itself—about a relation, does something "more unusal and difficult" than poets normally attempt. (All the power in these relations, all the wealth in James' lawn, will have to remain unanalyzed—not by James, who can't stop analyzing, but by Orr.)

Orr admits, though, that none of this meant anything to his father. Or perhaps better, he assumes, on the basis of what his father says to him about the poem, that its technical virtuosity meant nothing to him. For his father, the tent was interesting. It reminded him of tents pitched in open fields by traveling circuses, a sight he recalls from his own childhood. For Orr, though, the pleasure his father found in the poem wasn't adequate, for, he writes, "if reminscence was all that was needed, we could just as easily have been reading a magazine article about P. T. Barnum." He worries that what is properly poetic about Frost's lines are not a part of his father's enjoyment—the sound of the poem, its syntax, "the expert maneuvering that Frost does in order to unload the poem's only four-syllable word in the poem's penultimate line: 'In the *capriciousness* of summer air.'" Orr wanted his father to hear that—and to have it "help somehow" (Orr 2012, p. 192). But help with what? Orr doesn't tell us if the meter and rhythm of the poem made his father better able to bring intonation to his speech. That was, I thought, the point of the exercise. Given that Orr doesn't tell us whether reciting the poem helped his father's speech, I am going to assume that what served as the justification for the exercise was not its real agenda.

(Why can't a magazine article about P. T. Barnum be a poem?)

Instead of telling us about his father's voice, Orr tells us about his father's reminscences. And the circus, clearly, isn't sufficiently profound, not least because it doesn't seem to require poetic form for its elicitation. But is the clever use of syntax and metaphor in itself valuable? Should his father be taking the pleasure in the poem that Orr takes, and without that specific technical pleasure, is the poem a waste of time? Does poetry require formal difficulty—in its execution and in our appreciation of that execution—in order to be worthwhile? And do we have to be able to recognize, analyze, and describe these technical achievements in order to take pleasure in them?

And then there is the question of meaning. Frost's poem isn't that hard to understand—arguably its technical virtuosity is hidden by the relative simplicity of its meaning. Some readers of poetry, those who love secrets, might scoff at its lucidity. But there are secrets and there are secrets.

Does Orr suppose his readers know the poem? That they will go look it up? (Or is he hoping that they won't?)

Anyway, I did, in my copy of *The Complete Poems of Robert Frost* (1930; second printing 1949). (On the fly-leaf, "For Jimmie/21 Dec'50/Joe." From my dad to my mom, on her 29th birthday. There is a Sunday "Peanuts featuring 'Good ol' Charlie Brown" folded into fourths and tucked neatly into the front cover of the book. "Sally! Your beach ball is floating away. It's going clear across the lake!" "Stay calm, big brother … stay calm!" She addresses the ball: "**Okay, you stupid beach ball, come back here right now, or I'll see to it that you regret it for the rest of your life!**" The ball returns. "You have to know how to talk to a beach ball!" says Sally. Who thought this was so funny they needed to save it? What did it mean to them? There is a golf joke in there somewhere. And my brother Daniel.)

Okay, sorry, the poem.

She is as in a field a silken tent
At midday when a sunny summer breeze
Has dried the dew and all its ropes relent,
So that in guys it gently sways at ease,
And its supporting central cedar pole,
That is its pinnacle to heavenward
And signifies the sureness of the soul,
Seems to owe naught to any single cord,

> But strictly held by none, is loosely bound
> By countless silken ties of love and thought
> To everything on earth the compass round,
> And only by one's going slightly taut,
> In the capriciousness of summer air
> Is of the slightest bondage made aware. (Frost 1930, p. 443)

The "central cedar pole" that is the "pinnacle" of the tent, "to heavenward," "signifies the sureness of the soul," of her soul. When no particular string pulls, the pole and its tent seem free (although Frost does not use the word) despite—in fact because—it is "loosely bound/By countless silken ties of love and thought/To everything on earth the compass round." The more plentiful our ties, the more sure—the more upright and capacious—the soul. But when one pulls—the speaker of the poem? a chance summer breeze?—only then is it (but of course, the tent isn't aware, only "she" is), only then is *she* made aware of its/her bondage, however slight.

(After my brother Daniel died, my sister had a dream. She was late for church, standing in the back looking to see if she could find a seat. Daniel was there, in a full pew, with his daughers and me and my other brothers and my other sister and my sister's kids. Far too many for one pew. (When we were kids we could squeeze six, but that still meant two pews for the whole family. And endless fights over who had to sit with my mother.) But Daniel gestured to my sister to come and sit in the pew. She did and we were all there and there was more than enough room for everybody. That was my brother's heart.)

There is more in Frost's poem than might immediately catch one's eye.

(As my father lay dying, I read Tertullian's *On the Resurrection of the Flesh*. My father's edemic leg hung outside the bed clothes—my always so composed, so elegant, so immaculate father's body cast into disarray by pain and disease. My father, who, dead drunk, passed out with a cigarette so carefully balanced between the fingers of his right hand as it lay, folded gently on his left, across his chest, that only a perfect column of ash was left, hours later, when I found him, afraid to move the stub lest the ash scatter over his unburnt hand or shirt or the white sheet on which he lay.

My father, who, his leg run over, twice, by a drunken friend, got up, walked to the driver's side, and took the wheel.

My father, who caught me when I fell, blood running down his bright white shirt.

His swollen leg and foot fell out of the bed clothes. I wanted to grab hold and pull him back into this oh so painful flesh.)

"[P]oetry," Orr concludes, "needs a history with its readers." For Orr, poetry, to be useful, or helpful, or whatever it is he's seeking

> needs to have been read, and thought about, and excessively praised, and excessively scorned, and quoted in melodramatic fashion, and misremembered at dinner parties. It needs, in a particular and occasionally ridiculous way, to have been loved. If poetry could do nothing for my father that a thousand other things couldn't do, that was because it hadn't been a part of his life—just as when I'm eventually laid low, I will take little comfort in cello concertos and origami. (Orr 2012, pp. 192–193)

(So does a Christian need to have spent her life reciting the Psalms, being shaped by the poems' sound and their content, by the work that monks and nuns do with and to them, in order for their recitation as she dies to be "helpful"? Why then sing

Psalms over the beds of the dying, even those of the laity who may not have been particularly devout—although the Psalms were everywhere, and so everywhere heard, in the Middle Ages?)

(A friend lay with his eyes closed as the church choir sang around his bed. But what he seemed most to enjoy, or so I like to imagine, were the tidbits from a biography of Louis Zukovsky I was reading as we sat together in the late spring sun. He lay, with tubes attempting to stem the flow of waste that otherwise would come, periodically, from his mouth.)

Yet Orr knows that the story about technical expertise as a source of pleasure and distraction in the face of death is inadequate, because his father did respond, with intense pleasure, to at least one poem. But as Orr writes, "when he did so … it wasn't because of some rarity unearthed by the expertise of his clever son, or because of the uncanny genius of one of the definitive poems of our language" (Orr 2012, p. 193).

What Orr's father loved and enjoyed for a few of his dying days was Edward Lear's "The Owl and the Pussycat." (My friend loved this poem too, its divine foolishness.)

> The Owl and the Pussycat went to sea
>   In a beautiful pea green boat,
> They took some honey, and plenty of money,
>   Wrapped up in a five pound note.

Lear's nonsense poem gave Orr's father intense delight for a number of difficult weeks before the most difficult ones that lead to his death. Orr assumes that his father must have been familiar with the poem, or one like it; that the pleasure he took in it is tied to some dim memory of reading to his children many years before. "It was," Orr writes, "happy silliness, soon to end—and surely there were a hundred other things that might have given my father the same comfort—but this absurd poem was, in its own small way, *something*" (Orr 2012, p. 194). Orr cites the closing lines of the poem and his father's response to them:

> "Dear pig, are you willing to see for one shilling
>   Your ring?" Said the Piggy, "I will."
> So they took it away, and were married next day
>   By the Turkey who lives on the hill.
> They dined on mince, and slices of quince,
> Which they ate with a runcible spoon;
> And hand in hand, on the edge of the sand,
>   They danced by the light of the moon,
>   The moon,
>   The moon,
> They danced by the light of the moon.

> "I really like," said Dad, "the runcible spoon." Reader, there are worse things to like. Or to love. (Orr 2012, p. 194)

In *Infinite Jest*, a novel I've never read, David Foster Wallace has his character, Remy Marathe, a Quebecoise terrorist, say, "Choose with care. You are what you love. No?" Truth or truism, this is just the sort of literary pronouncement David Orr finds deeply embarassing. So do I, a lot of the time, when writers proclaim that poetry and fiction can radicalize consciousness, transform the political sphere, recreate the everyday world, provide the sole and necessary grounds for ethics, usher us into an easeful death—I cringe. Poetry and fiction are marks on the page and sounds. And death, like literature, is difficult.

(When my father and mother, my brothers and sister and friends died, I wanted to read difficult books. Really really difficult. Only the incomprehensible was comforting in the face of how hard death is.)

Who knows why Orr's father liked Lear better than Frost, although I have a feeling it has less to do with prior knowledge of the poem than with its ability to help him articulate sound and with it, perhaps, feeling—feelings of whimsy and humor. (My sister was in a fever as she died. "Better get used to it," someone said. We laughed and laughed. My sister, I fear, was unconscious.) Sounds elicit whole worlds of feeling. They are among the strongest—and arguably the last—of the strings that hold us to earth and enable us to reach … heavenward? (My sister and I sang to Daniel, snatches of half-remembered songs, as he gasped out his last, morphine slowed breaths.) A novel by Henry James or David Foster Wallace, a poem by Robert Frost or Edward Lear, a Beethhoven violin concerto or a Dum-Dum Girls song, it's the saying, the listening, the reading, the repetition, the way something we love pulls us back to earth even as it lets us fly—that's what I love. (In my [day] dreams, it's what I am.)

(In my [day]dreams, difficulty's reverberations sustain and shake and yet still somehow soothe.)

***

Someone close to me said that he doesn't care what goes on at his funeral, except that he wants loud music. Really really loud.

What's the aging hipster's version of the Psalms?

> Oh it's a game, hold tight
> Can you shut your eyes?
> Shut out the light
> Death is so bright
>
> From dreams you wake to shock
> To find its true
> But she's not you
> No she's not you
>
> And you'd do anything to bring her back
> Yes you'd do anything to bring her back
>
> Oh I wish it wasn't true
> But there's nothing I can do
> Except hold your hand
> Except hold your hand
> 'Til the very end (Gundred 2011)

The words are good, but it's the wall of sound, the layers of feedback, the echoing resonanting chaos of it all, that bring the house down.

References

Frost, Robert. 1930. *The complete poems of Robert Frost*. New York: Henry Holt and Company.
Gundred, Kristin. 2011. *Hold your hand*. London: Kobalt Music Publishing Ltd.
Howe, Fanny. 2003. *Tis of thee*. Berkeley: Atelos.
Howe, Susan. 2010. *That this*. New York: New Directions.
James, Henry. 1975. In *The portrait of a lady*, ed. Robert D. Bamberg. New York: Norton.
Orr, David. 2012. *Beautiful & pointless: A guide to modern poetry*. New York: Harper.

Chapter 4
Negotiating Advance Directives in a Navajo Context

Michelene Pesantubbee

Abstract The introduction of Advance Directives in the second half of the twentieth century allows patients to make end-of-life decisions while they are mentally and emotionally fit to do so. However, despite the efforts of health care professionals to get Americans to prepare Advance Directives, the response of Americans has been disappointing, even more so among Native Americans. This essay explains how Navajo traditionally conceive of death and dying and how Advance Directives and medical practices clashed with their conceptions. In an effort to encourage Navajo patients to complete Advance Directives at Fort Defiance Hospital medical personnel participated in cultural-sensitivity training and introduced Navajo-centered approaches for documenting end-of-life decisions and supportive care. This summary of the approach taken at Fort Defiance Hospital demonstrates the success of using culturally sensitive accommodations.

Any examination of death and dying issues among Native Americans presents challenges about whom to include or exclude. According to the 2010 United States census, 5.2 million people identified themselves as being of Native American descent, though 78% of them live outside of Native American or Alaskan Native communities and therefore have limited opportunities to participate in their tribal traditions (Norris et al. 2012, pp. 1, 12, 20). Among those people who did grow up in tribal communities, specifically reservations, very few spent time studying their traditional religious ways. Unlike religious groups that have written sacred texts and organized structures of leadership, Native Americans do not have systematic courses of study, systems of accreditation, or formal discussions of the nature of God or beliefs. One learns by observation or instruction from a local specialist, and one cannot do that, at least not effectively, without spending extended time in the community. It is also sometimes difficult to separate Native American traditions from Christian ones. All Native American communities have been exposed to Christianity

M. Pesantubbee (✉)
University of Iowa, Iowa City, IA, USA
e-mail: michelene-pesantubbee@uiowa.edu

T. D. Knepper et al. (eds.), *Death and Dying*, Comparative Philosophy of Religion 2, https://doi.org/10.1007/978-3-030-19300-3_4

for at least 300 years and in some cases for more than 500 years. Many Native Americans identify as Christian, others identify as traditional and Christian, and among those who identify as traditional, some engage in intertribal traditions or in the traditions of a tribal community other than their ancestral one. The legacy of colonization and missionization as well as the inclusion of hundreds of distinct indigenous religious traditions make any study of a general Native American response to death and dying, or in this case Advance Directives, impossible. Heterogeneity is also true of any single Native American community.

However, with these concerns in mind, this essay will offer one perspective on advance-directive experiences of one indigenous group—the Navajo. The Navajo were selected for several reasons. First, the size and location of the Navajo Reservation in the Southwest have helped the Navajo to retain their language and culture to a much greater degree than many other native nations. At least half the population still speaks fluent Navajo (Siebens and Julian 2011, p. 2), and the percentage of Navajo who participate in traditional Navajo ways is relatively high compared to many other Native communities in the U.S. Second, the Navajo Nation is the second largest Native American nation in the United States with a population of 332,129, and 50% of Navajo live on the reservation, a much higher percentage than we find among many other tribes (Norris et al. 2012, p. 17). The Navajo Reservation also has the largest number of people identifying with one tribe only (Norris et al. 2012, p. 18). The high level of cultural retention among Navajo offers us a good chance of assessing the effects of medicalization on traditional beliefs and practices, at least of some Navajo in particular. Third, some targeted programs on the Navajo reservation have led to some unexpectedly high rates of participation. This change in attitude among Navajo is intriguing because Navajo are considered reticent to talk about end-of-life issues, and they have a history of not trusting hospitals. This raises the question of what changed Navajo minds about preparing Advance Directives.

Before addressing specifics of Navajo Advance Directives, qualifying remarks are in order. First, although Navajo recognize a general version of their origin story and related ceremonies, details or attributes vary from person to person. Navajo religious specialists (*hataali*), commonly referred to as singers or chanters, learn their vocation typically from someone in their family or in the community. The practice of handing down songs, prayers, medicine bundles, and rituals means that although there is a recognized format and content for specific ceremonies, variations do exist. Therefore, this representation of Navajo experiences is based on whose work I have read and which people I have talked to, and it therefore does not represent all Navajo perspectives. It is also important to point out that I am not Navajo, so I offer an outsider's understanding of Navajo religious traditions.

Second, outsiders often mischaracterize Navajo ideas about death. Some people portray them as dreadfully afraid of death. In fact, Navajos do not have an unreasonable fear of death, but rather, as Gary Witherspoon puts it, they have a "tremendous respect for life" (1977, p. 20). Just like the rest of us, they do not look forward to dying, though they do understand that death is part of the natural cycle of life. The confusion may arise from the fact that traditionally, Navajo avoid contact with the dead or talking about a deceased person, again not because they are afraid of death, but because the ghost or spirit of the dead may still remain near the physical remains and cause harm to those who ignore customary restrictions (Witherspoon 1977, pp. 19–21, 205 n. 3).

When we think of medicalization of death and dying, Advance Directives or end-of-life directives do not immediately come to mind. The Advance Directive is a fairly recent phenomenon in the United States. Physicians, insurance agents, social workers, and retirement specialists invest time in educating clients about Advance Directives and encouraging them to complete Advance Directives. The concept of *living wills*, which allows patients to instruct physicians on what life-sustaining measures they want, was first introduced in 1967. Not long after that health professionals began advocating for durable powers of attorney in which a patient can designate someone else to make medical decisions on their behalf. Within 30 years every state had enacted a policy. By the early 1990s, states began passing Do Not Resuscitate (DNR) legislation (Benson and Aldrich 2012, pp. 11–12). In spite of the health care professions' efforts to encourage Americans to prepare end-of-life directives, their success rate has been disappointing, even more so among Native Americans. A study of Navajo experiences with Advance Directives provides some insight into not only the difficulties of encouraging Native American acceptance of Advance Directives but also approaches to providing a culturally sensitive introduction to Advance Directives.

To understand Navajo reticence and acceptance of Advance Directives we must first have some understanding of Navajo concepts of life and death. For Navajos, air, or what some call wind, is equated with life. They identify seven classes of wind souls. One called the "small wind" or "little inner wind" enters the fetus once the fetus has developed into human form. If you have ever wondered what enables the fetus to kick and move around at the most inopportune times, the Navajo would say it is the small wind. The small wind enables the fetus to move and controls its digestive system. After birth, the small wind remains with the baby and helps it to continue to grow. However, upon birth the child acquires a second wind, called the in-standing wind soul. The in-standing wind soul is understood to control the functioning of the brain, which in turn, controls a person's thoughts and bodily movements (Witherspoon 1977, pp. 30–31). What we can gather about the relationship of air and life according to Navajo thought is that air makes life possible and life is manifested in movement. To quote Gary Witherspoon, "Movement is the basis of life, and life is exemplified by movement" (1977, pp. 53). Since air or the breath of life is the source or power of movement as evidenced by the little wind and the in-standing wind soul, when the air leaves the body, movement ceases to occur, and life ends (Ladd 1957, p. 417; Witherspoon 1977, pp. 20–21). But why must movement cease and life end? Why does the wind leave the body? The answer for the Navajo lies in their origin story.

In their story it is told that First Man and First Woman originally wanted people to live forever. However, the Sun told the people that as payment for the light he provides, he would claim lives. When First Man and First Woman saw deceased people in the below world, it made them unhappy and they decided to try to change things. So they threw "a log into a lake, saying that if it floats, people will live forever, but if it sinks, people will die." About that time ole Coyote came along and proceeded to toss a stone into the water declaring that if the stone doesn't float, then there will be death. Of course, the stone sank (Reichard 1950, p. 42; Levy 1998, p. 66). The people were upset but Coyote explained to them that "death had to be a part of the scheme of this world…. If death did not occur, the earth would soon be overcrowded, and there would be no room for corn fields" or for future generations. He told the people, "It is better that each one of us should live for a limited time on this earth and then leave and make room for the children" (Reichard 1950, p. 42). From this story, the Navajo understand that death is not an evil willed upon the world by Coyote but rather more an unavoidable necessity (Witherspoon 1977, p. 19; Reichard 1950, p. 42; Levy 1998, p. 112).

If death is an unavoidable necessity, then one must consider how and when death should occur. For Navajo, a natural and highly desirable death is one that results from old age. This reasoning derives in part from Navajo understanding that birth and death are structural opposites; "one cannot exist without the other" (Witherspoon 1977, p. 20). Witherspoon explains this concept as follows, "life is considered to be a cycle which reaches its natural conclusion in death of old age, and is renewed in each birth. Death occurs so that birth is possible" (Witherspoon 1977, p. 20; Wyman 1970, p. 573). Navajo tell of a time when one of their culture heroes, Monster Slayer, tried to kill Old Age. Old Age spoke up and said, "In spite of all, I am going to live on, my grandchild, … should you kill me dying would cease…. Then too giving birth would cease, … and this present number of people would continue in the same amount for all time to come" (Witherspoon 1977, p. 20; Wyman 1970, p. 573). In other words, the dynamic flux of the universe, that motion which continually recreates order, would cease and so too would the world (Hoijer 1964, p. 146; from Griffin-Pierce 1992, p. 24). This made sense to the Navajo because they witnessed this pattern of motion throughout the universe. In spring plants bloom; in winter they die. The sun rises; the sun sets. If plants quit growing and the sun quits rising, the world will end. So too must birth and death continue in order for the life cycle to continue.

If a desirable death is one that occurs from old age then how do Navajo account for or explain death by any other means? For Navajo, death before old age is considered unnatural and tragic because it prevents the natural completion of the life cycle (Witherspoon 1977, p. 20). Premature death is believed to result from malevolent intentions and deeds. Throughout one's life, a Navajo strives to live according to *hozho*, a life of beauty, harmony, and happiness. However, malevolent forces exist, and if they are not discovered and transformed into benevolent ones, death can occur before old age is reached (Witherspoon 1977, p. 35). One of many ways in which malevolent forces can be manifested or ascertained is through the presence of illness.

We all know illness or injury have physical causes and manifest symptoms. The question for Navajo is what brought about those physical causes? Navajo understand the physical manifestation of illness to be due to impaired relationships or states of imbalance. Their creation stories teach them that they should maintain proper spiritual relationships with the *diyin dine'é* or holy people; they must properly carry out their responsibilities to the environment, execute proper ceremonial practices, and maintain harmony with each other. When Navajo fail to maintain proper relationships and attitudes, illness results. Illness then is the result of behavior that leads to disharmony (Gill 1987, p. 122, 1992, pp. 76–77). As Griffin-Pierce explains, when people engage in acts that disrupt the balance in the forces of nature, they "may be made ill by the forces thereby unleashed" (Griffin-Pierce 1992, p. 35). If the holy people are offended, they may direct life-threatening influences towards the offending person, leading to a state of imbalance or malevolence. "The illness is overcome" when balance or *hozho* is restored through ceremony (Gill 1992, p. 77). As Jennie Joe and her co-authors explain, the central aim of all Navajo healing ceremonies is not to cure the physical ailments, "but rather to empower the patient so he/she can work towards restoring balance or harmony of mind, body, and spirit" (Joe et al. 2016, p. 35).

Navajo engage in ceremony in order to reestablish proper relationships with the holy people and to compel them to remove the spell causing the illness. The prayer obligates the holy being to remove the spell and allows the person to return to a state of health (Gill 1987, pp. 119, 122). In this excerpt from a Holyway Prayer we see how prayer works:

> This very day you must take your spell out of me by which you are bothering me;
> This very day you have removed your spell from me by which you were bothering me.
> (Gill 1987, pp. 119–20)

After an obligatory relationship is established, the holy being is beseeched to take the illness out: "You must take your spell out." In the second line, the act of recovery is then initiated: "You have removed your spell." If the prayer is properly executed, then the holy person should fulfill his/her obligation by removing the spell (Gill 1987, p. 25).

After going through the ceremonial process, an ill person may still die. One might assume in such a case that Navajo would lose faith in the ceremony, but one would be wrong. Ceremonial response to illness is not questioned in such cases, but rather the diagnosis, location or timing of the ceremony, or the performance of the ceremony are questioned. Sam Gill explains that if the diagnosis is incorrect, then the wrong ceremonial was probably performed or the ceremony was inadequately performed. Perhaps the chanter did not perform the ceremony in a properly consecrated enclosure or during the appropriate season of the year. In other words, if the ceremony is not carried out "exactly as required by conventions" or the participants do not have the proper intentions or adhere to applicable restrictions, then the ceremony will fail to cure the person (Gill 1987, pp. 124–125).

If a ceremony does not cure the ill person the first time, Navajo may repeat the ceremony several times over a period of time. If the ceremony continues to be ineffective, then as a last resort, Navajo might take the ill person to a hospital. By waiting for the proper season or time and repeating the ceremony several times, a Navajo might not be taken to the hospital until their illness is quite advanced and the patient's chances of survival low. This delayed action contributed to Navajo perceptions that hospitals are "houses of death." So many people died in hospitals that Navajos began to say that no one could get well in places so polluted by the winds or *chįįdii*, ghosts of the dead (Levy 1978, p. 401).

The high rates of death in hospitals resulted from more than just delayed hospitalization. In an article by the National Cancer Institute (NCI) they noted that although as a group Native Americans/Alaskan Natives have lower rates of most types of cancer than non-Hispanic white Americans, "they are more likely to be diagnosed with late-stage disease, and their cancer survival is generally poorer than that of other groups" (NCI 2011). Before the advent of antibiotics, Navajos suffered high fatality rates from upper respiratory infections and tuberculosis. It is no wonder that Navajo associated hospitals with death (Levy 1978, pp. 401–402).

Navajo distrust of hospitals also stemmed from the way in which hospital personnel responded to illness. The difference between Navajo and western responses to illness is clearly seen in an event that Witherspoon observed on the Navajo reservation in 1969. In his book, *Language and Art in the Navajo Universe*, he tells a story about an 82-year old woman who suddenly became ill. She was quickly taken to a hospital, where she fell into a coma and the hospital kept her alive by intravenous feeding. The family decided to consult a diviner who diagnosed her case as being caused by contact with the ghost of a deceased non-Navajo and prescribed the three-day Enemyway rite. When the family informed the doctors that they wanted her released so that the Enemyway rite could be performed, the doctors tried to stop them by telling the family that if she was taken off intravenous feeding, "she would soon die." Despite the risk, the family decided to take her home anyway and had the rite performed. Before the end of 3 days, she regained consciousness and began eating and talking. Not long after, she was up and walking about (1977, pp. 13–14). In this particular case, what we see is that the doctors had little confidence in Navajo healing ceremonies that focused on restoring *hozho*, and the Navajo had little faith in western medicine that concerned itself with physical symptoms and physiological causes rather than curing the disease by restoring balance (Levy 1978, p. 402). Neither understood the other's concern about how to care for someone facing death.

This lack of understanding between western and traditional healers is not unusual. During their early years, Indian Health Service (IHS) and other hospitals serving Native Americans did not always welcome or accommodate the services of traditional tribal practitioners. Nor did they try to find ways to address Navajo concerns. However, in the 1990s attitudes began to change. The federal government gradually began to recognize the importance of cultural traditions in delivering health care of Native Americans, and in 1994 "then-director of the IHS, Michael H. Trujillo, a Native American physician, issued an administrative initiative asking all regional IHS facilities to find ways to work more closely with local Native practitioners" (Joe et al. 2016, p. 31). Six years later Chinle Comprehensive Health Care Facility officially opened their Office of Native Medicine (ONM) with the hiring of the first traditional Navajo practitioner (Joe et al. 2016, p. 31). One of the objectives of ONM was to provide cultural-sensitivity training to health care providers by providing educational programs about Navajo medicine, healing, culture, and language (Joe et al. 2016, p. 32).

Ironically, by providing cultural-sensitivity training to hospital staff, ONM reinforced staff resistance to discussions of end-of-life decisions with Navajo patients and their families. As Dr. Timothy Domer, the Director of the home-based care program at Fort Defiance Hospital explained to *New York Times* writer Ben Daitz, as medical personnel became aware of Navajo beliefs about death and the idea that talking about it is thought to bring it about, they did not want to raise the topic of end-of-life directives with Navajo patients. They avoided discussing preparation for death with the family. They didn't want to offend Navajo, and they didn't want to risk Navajo avoidance of hospitals for fear of causing death of their relatives. Yet, hospital staff needed to bring up the subject because Navajo were bringing in dying patients, "particularly elderly patients who spoke only Navajo" (Daitz 2011).

Not only did Navajo tend to delay taking their gravely ill relatives to the hospital, they also started taking their dying relatives to the hospital in order to avoid some of the restrictions associated with the care of the deceased. Traditionally, when someone was gravely ill, several close members of the family and the healers stayed with sick person until they were certain death was near. Then everyone left except for one or two people who performed the necessary rites (Griffen 1978, p. 367; from Reichard 1928, p. 141). The family also made sure that outsiders were warned about the impending death, so that then and later they would stay away from the area (Griffen 1978, p. 368). After the person died, the hogan, or traditional Navajo house, "in which death occurred was burned or abandoned and mourners spent the four days of seclusion in another dwelling" (Griffen 1978, p. 374). By taking the dying to the hospital Navajo avoided having to abandon the hogan where the person died. As Dr. Domer explained, many Navajo are too poor to afford to relocate and build another hogan (Daitz 2011).

All of these factors together contributed to Navajo perceptions that hospitals are places to take those nearing death. Hospitals became places to die, and yet Navajo were reluctant to talk about death with the very people they had entrusted with their loved ones. According to a 1995 study, 86% of 22 Navajo interviewed considered advance-care planning "a dangerous violation of traditional Navajo values and ways of thinking" (Carrese and Rhodes 1995, p. 828). Health care providers and Navajo healers needed to find ways to accommodate and integrate the practices of the other in order to meet the needs of Navajo facing end-of-life decisions in IHS hospitals.

A shift in attitude among some traditional Navajo arguably began in 2006. That was the year Dr. Domer, a geriatrician who practiced medicine on the Navajo reservation for years, started a home-based care program. When he reviewed the hospital records of Navajo patients at Fort Defiance Indian Hospital in northeastern Arizona, he found zero Advance Directives in their charts. The absence of health directives in Navajo medical charts is consistent with what a 2005 NCI-funded study of IHS tribal health directors found, which was that a majority of Native American communities were not receiving advanced-care planning, bereavement support, care for the dying, or hospice care (National Cancer Institute 2011; Daitz 2011). Dr. Domer wanted to find a way to work with his terminal patients and their families and change the way dying Navajo live.

But first he had to find a way to sensitively address Navajo ideas about death and medical providers' perception that Navajo are unwilling to discuss end-of-life issues. Health care providers had long held that Navajo elders "will not openly discuss issues related to death and dying" (Baldridge 2011, p. 1). In the 1995 Carrese study three of the Navajo interviewed were trained as biomedical health care providers. Although they acknowledged their responsibilities to work on living wills, they also noted that as Navajo, one did not discuss end of life issues (Carrese and Rhodes 1995, p. 828). The authors of the study reported that "discussing negative information"—such as end-of-life care—was considered culturally offensive and "potentially harmful" (Carrese and Rhodes 1995, p. 826, Baldridge 2011, p. 2). In addition, the IHS manual clearly stated that medical personnel must respect patients' tribal customs and traditional beliefs that relate to death and dying whenever possible. However, an individual patient's preference and beliefs takes precedence over his or her tribal customs and traditional beliefs (U. S. Dept. of Health and Human Services 2005).

Dr. Domer needed to find a way to get his staff to accept the charge of discussing end-of-life issues with Navajo patients and their families. This was not an easy task. He held several meetings to discuss the issue. As Dr. Domer recalled, "Nine years ago we weren't having much success. We came out of end-of-life conversations feeling drained and frustrated, and our staff wasn't comfortable." He explained, "The biggest problem that we have [to] overcome is the reluctance of the staff to bring up the issues with the patients." So he asked "Navajo social workers to come up with a plan that would use words and images that were culturally acceptable for end-of-life discussions—and they did" (Baldridge 2011, p. 3).

Since broaching the subject of death was frowned upon, Mrs. Begay, the cross-cultural coordinator for the home-based care program at Fort Defiance Indian Hospital and her staff, came up with an approach to opening the doors to discussing Living Wills and Advance Directives with Navajo patients. They created a poem, part of which goes as follows: "'When that time comes, when my last breath leaves me, I choose to die in peace to meet *Shi'dy'in*'—the creator" (Daitz 2011). This poem is consistent with Navajo beliefs and opened the door to further discussion. According to Dr. Domer, 90% of the patients in the home-based care program signed the poem and other standard directives. Once the staff became comfortable with the subject and knew the right words to use, he found the patients were quite often willing and "often eager to have these discussions and formally establish their Advance Directives and durable medical powers of attorney" (Baldridge 2011, p. 5).

One reason the poem works is that it is a directive of the Navajo's choosing. Advance Directives in some ways are consistent with Navajo traditional practices. Some older Navajo will distribute some of their personal property to relatives in anticipation of death. They may announce their wishes in front of a gathering of relatives to ensure their wishes are carried out. Some might leave sheep permits to particular relatives; others might distribute jewelry or even songs or medicine bundles. Some traditional Navajos have been known to ask a missionary or another white person to prepare written wills. They may let their relatives know how they want their body prepared for burial (Shepardson 1978, p. 389). In other words, Navajo already employed verbal directives to make sure their wishes are known and the poem served to make the connection between traditional Navajo directives and state health directives and Navajo concepts of death.

Although the program is considered successful, Advance Directives raised other cultural issues for Navajos. In a case study on patient-centered care at Chinle Comprehensive Healthcare Facility, Jennie R. Joe and her co-authors related the following story:

> A 45-year [old] patient on dialysis wanted to discontinue this procedure against the family's wishes. As the patient's health was failing, the family requested the help of ONM [Office of Native Medicine] to conduct a minor ceremony, but because the patient had signed an advance directive, the practitioners had to deny the request. (Joe et al. 2016, p. 40)

Why is this a problem? One, Navajo traditional practitioners tend to work with both patients and their family members. The patient and family members may disagree but the family members discuss the steps that should be taken on behalf of the patient. That is part of their responsibility as relatives. End-of-life directives do not allow for consideration of family wishes. Two, Navajo traditional practitioners must establish clan relationships. If the practitioner has a kinship connection to the patient, then certain rules come into play including proper etiquette and reciprocal kinship obligations that can challenge federal rules about patient confidentiality (Joe et al. 2016, pp. 40–41). Three, another possible conflict arises if the patient or family requests a traditional diagnosis and determination of relevant interventions or ceremonies. Preparation for ceremonies as well as the ceremony itself takes considerable time (Joe et al. 2016, p. 41), which may interfere with hospital treatment or schedules.

At Chinle the hospital took steps to minimize some of these issues by constructing a hogan on the hospital grounds, which Navajo traditional practitioners could use for carrying out brief ceremonies for patients and their families. In one wing of the hospital they created a circular healing room shaped like the inside of a hogan. ONM staff used this room "for patient consultation or for minor ceremonies when a patient is too ill to leave the hospital" (Joe et al. 2016, p. 34). They also built two sweat lodges, "one for men and one for women" (Joe et al. 2016, p. 34). By providing such facilities, Chinle Hospital has reduced the amount of time needed for preparation and travel, and the patient remains near medical facilities. By providing hogan-like facilities, elder patients can experience the hospital as a place where the cause of their illness is treated and not as a place of death.

In summary, Navajo willingness to discuss end-of-life directives benefitted from cultural-sensitivity training on the part of health professionals, accommodation for traditional healing-ceremonies, and a change of attitude among Navajo toward physicians, who they no longer saw as purveyors of death, but as healers who can provide topic treatment in concert with Navajo traditional healing practices. By incorporating culturally sensitive accommodations for Navajo, Navajo participation in end-of-life directives went from 0 to a 90% rate at Fort Defiance Indian Hospital (Baldridge 2011, p. 5, Daitz 2011). Although signing end-of-life directives raises issues of kinship responsibilities, signing them no longer indicates rejection of Navajo belief in traditional medicine, but rather participation in a bicultural approach to treating symptoms along with causes.

References

Baldridge, Dave. 2011. Moving beyond paradigm paralysis: American Indian end-of-life care. National Association of Chronic Disease Directors: Critical issue briefs. www.chronicdisease.org. Accessed 19 Feb 2017.

Benson, W. F. and N. Aldrich. 2012. Advance care planning: Ensuring your wishes are known and honored if you are unable to speak for yourself. Centers for Disease Control and Prevention: Critical issue brief. www.cdc.gov/aging. Accessed 21 Feb 2017.

Carrese, Joseph A., and Lorna A. Rhodes. 1995. Western bioethics on the Navajo reservation; benefit or harm. *Journal of the American Medical Association* 274 (10): 826–829. https://doi.org/10.1001/jama.1995.03530100066036.

Daitz, Ben. 2011. With poem, broaching the topic of death. *The New York Times*, January 24.

Gill, Sam. 1987. Prayer as person: The performative force in Navajo prayer acts. In *Native American religious action*, 113–128. Columbia: University of South Carolina Press.

———. 1992. It's where you put your eyes. In *I become part of it*, ed. D.M. Dooling and Paul Jordan-Smith, 75–87. New York: HarperCollins Pub.

Griffen, Joyce. 1978. Variations on a rite of passage: Some recent Navajo funerals. *American Indian Quarterly* 4 (4): 367–381.

Griffin-Pierce, Trudy. 1992. *Earth is my mother, sky is my father*. Albuquerque: University of New Mexico Press.

Hoijer, Harry. 1964. Cultural implications of some Navajo linguistic categories. In *Language in culture and society*, ed. D. Hymes, 142–153. New York: Harper & Row.

Joe, Jennie R., Robert S. Young, Jill Moses, Ursula Knoki-Wilson, and Johnson Dennison. 2016. At the bedside: Traditional Navajo practitioners in a patient-centered health care model. *American Indian and Native Mental Health Research* 23 (2): 28–29. https://doi.org/10.5820/aian.2302.28.

Ladd, John. 1957. *The structure of a moral code*. Cambridge, MA: Harvard University Press.

Levy, Jerrold E. 1978. Changing burial practices of the Western Navajo: A consideration of the relationship between attitudes and behavior. *American Indian Quarterly* 4 (4): 397–405.

———. 1998. *In the beginning: The Navajo genesis*. Berkeley: University of California Press.

National Cancer Institute. 2011. A look at end-of-life care issues for Native Americans. *Lifelines from the National Cancer Institute*. Reprinted in *Indian Country Today*.

Norris, Tina, Paula L. Vines, Elizabeth M. Hoeffel. 2012. The American Indian and Alaska Native population: 2010. U.S. Census Bureau. https://www.census.gov/library/publications/2012/dec/c2010br-10.html. Accessed 19 Feb 2017.

Reichard, Gladys. 1928. *Social life of the Navajo Indians with some attention of minor ceremonies. Columbia contributions to anthropology* 7. New York: Columbia University Press.

———. 1950. *Navajo religion: A study of symbolism*. Princeton: Princeton University Press.

Shepardson, Mary. 1978. Changes in Navajo mortuary practices and beliefs. *American Indian Quarterly* 4 (4): 383–395.

Siebens, Julie, and Tiffany Julian. 2011. Native North American languages spoken at home in the United States and Puerto Rico, 2006–2010. United States Census Bureau, Report No. ACSBR/10-10. https://www.census.gov/library/publications/2011/acs/acsbr10-10.html. Accessed 18 Feb 2017.

U.S. Department of Health and Human Services. 2005. *Indian health manual, part 3—professional services*, Chapter 26—"Patient Self-Determination and Advance Directives," 4 January. https://www.ihs.gov/ihm/pc/. Accessed 19 Feb 2017.

Witherspoon, Gary. 1977. *Language and art in the Navajo universe*. Ann Arbor: The University of Michigan Press.

Worldatlas. n.d. Biggest Indian reservations in the United States. https://www.worldatlas.com/articles/biggest-indian-reservations-in-the-united-states.html. Accessed 17 Feb 2017.

Wyman, Leland C. 1970. *Blessingway*. Tucson: University of Arizona Press.

Chapter 5
The Cult of Santa Muerte: Migration, Marginalization, and Medicalization

Eduardo González Velázquez, Eduardo García-Villada, and Timothy D Knepper

Abstract The 2015–2017 programming cycle of The Comparison Project hosted two lectures on Santa Muerte: the first by Prof. Eduardo González Velázquez, a professor of history at Monterrey Tec National School of Social Sciences and Government (Guadalajara, Mexico), an international partner of Drake University; the second by Prof. Eduardo García-Villada, a professor of Spanish at Drake University. In this essay, we weave these two lectures into a composite picture of the role of the cult of Santa Muerte with regard to death and dying. Section one draws on Prof. García-Villada's initial encounter of the cult and consequent exploration of its history and practices. Section two contains the majority of Prof. González's lecture about how the cult serves migrants and other marginalized sectors of Central American society. Section three is an excerpt of Prof. García-Villada's lecture on what the prayers to Santa Muerte reveal about her nature and the function of her cult vis-à-vis traditional power structures. Finally, section four contains the reflections of The Comparison Project's director, Tim Knepper, about the relationship between the cult of Santa Muerte and the theme of the 2015–2017 programming cycle: medicalized dying. In a time when fear of death is a constant presence, the devotees of Santa Muerte have turned to her as a saint, if not a deity, who can protect and provide for them in a way that the state and church cannot, without regard for how marginalized or deviant they might be. In this sense, the cult of Santa Muerte serves an end not unlike that of bioethics: the attempt to control death.

E. G. Velázquez (✉)
Monterrey Tec National School of Social Sciences and Government, Guadalajara, Mexico

E. García-Villada · T. D. Knepper
Drake University, Des Moines, IA, USA
e-mail: eduardo.garcia@drake.edu; tim.knepper@drake.edu

T. D. Knepper et al. (eds.), *Death and Dying*, Comparative Philosophy of Religion 2, https://doi.org/10.1007/978-3-030-19300-3_5

5.1 Encountering Santa Muerte

It was the summer of 2010 when I, Eduardo García-Villada, had my first encounter with Santa Muerte. In the heart of the historical center of Mexico City, on the corner of Emiliano Zapata and Jesús María, a mere four blocks from the Metropolitan Cathedral and National Palace, there stood a life-size statue of a skeleton. It was dressed with a mantle, like any statue of the Virgin Mary found in Latin American Roman Catholic churches, which people in Mexico call *bultos* ("bundles"). The skeleton was on top of a table, with a collection of offerings laid at her feet. The *bulto* was unattended. No one was paying attention to it at the time, and there wasn't anyone nearby to ask about it. My reaction was one of shock. I didn't know what that dressed skeleton represented, though its allusion to the Virgin Mary was unavoidable. Such was my first encounter of Santa Muerte.

Not long afterwards, I began reviewing the seemingly countless references to death in Mexico: the recently excavated Tzonpantli Altar of the Templo Mayor, adorned with endless skulls; the Aztec goddess Coatlicue, whose face is a deadly double snake and whose skirt is made of snakes and skulls; Mesoamerican myths of the underworld written in codices and stones; and more. Did a contemporary manifestation of death like the skeleton saint have origins this far back in time?

Research in the field of Mexican anthropology and archaeology has found evidence of a cult of death in the ancient history of Mexican indigenous people (León-Portilla 1993; León 2007; Pennock 2012). The Aztecs in particular were well-known for their sacrificial practices and ostensible obsession with death. Their pantheon includes gods and entities that were associated with the dead, some of whom were female. For example, there are the *Cihuateteo*, the "Divine Women" who died in labor and went directly to a better place; *Coatlicue* ("Skirt of Snakes"), the earth mother goddess who is associated with sacrifice and death; and *Mictlancihuatl* ("Lady Death"), who along with *Mictlantecuhtli* ("Lord Death"), guards the underworld of *Mictlan* (Kroger and Granziera 2012, pp. 179–180, 184–187, 199–200).

With the Spanish conquest of Mexico, indigenous Mexican and Spanish Catholic views about death began to fuse. Perdigón (2008) and Chesnut (2012) trace the origin of Santa Muerte devotion to a syncretism between medieval Catholicism and Aztec religion. But it is not until the end of the eighteenth century that the first references to the veneration of Santa Muerte appear in the center of New Spain. And it is not until the latter half of the twentieth century that Santa Muerte devotion comes out into the open. Although some references identify the state of Hidalgo, where today they venerate the Black Angel, as the place where Santa Muerte worship began, in reality her cult found support in various states of the Republic: Guerrero, Veracruz, Tamaulipas, Campeche, Morelos, Jalisco, Estado de México, Sinaloa, and Mexico City.

The origins of Santa Muerte devotion therefore remain opaque. For some, its beginnings are pre-colonial, or in the very least the result of a combination of the pre-Hispanic and the colonial. For others, the "inspiration" for the cult is closely linked to the history of the practices and doctrine of the Catholic church and its preaching about the "Good Death." And so, controversy about the cult's origin continues.

Nevertheless, the main contours of the cult have taken shape. Santa Muerte is herself a many-named folk "saint." "The White Girl," "The Great Saint," "The White Lady," "The Mother," "The Beautiful," "The Skinny," "The Holy Girl," "The Godmother," "The Black Girl," "The Patron," "The Girlfriend, Boss," "The Lady of the Scythe," "The Death," and "The Cold" are among these many names. She is represented as a feminine figure dressed with a long, white satin tunic that leaves only her emaciated face and skeletal hands uncovered. She wears a halo of gold around her head, carrying in her right hand a scythe (with which to "harvest souls"), a rosary, and a scale (to represent justice), and in her left hand a globe and an hourglass. She is petitioned with different colored candles: black, for protection and strength; red, for love and passion; white, for protection for the family and to cleanse negative energies; blue, for protection for students; orange, to obtain money; green, to solve legal problems and maintain unity for loved ones; purple, for health; brown, to invoke the High Spirit; yellow, for good luck; beige, for peace and harmony for the home and business; and amber, to heal drug addicts and alcoholics.[1] Devotees also bring her gifts such as flowers, candy, fruit, liquor, water, and cigarettes.

These devotees often come from the margins of society: prisoners, prostitutes, homosexuals, transvestites, criminals, outcasts, and the impoverished and disenfranchised. It is no coincidence that the place where Eduardo García-Villada first encountered the statue of Santa Muerte is nearby the Barrio Tepito neighborhood, the epicenter of the Santa Muerte devotion in Mexico City. Barrio Tepito is known as a rough place that is home to open-air markets selling counterfeit goods, narco culture, and those who deviate from the supposed economic, political, cultural, and religious norms of the country. In this neighborhood, a woman named Enriqueta Romero has kept a shrine to Santa Muerte since 2001. She organizes well-attended prayer sessions for reciting the rosary, and is in fact credited as the source for the prayers that are used at the shrine (Roush 2012).

"The Skinny Woman" is eclectic, syncretic, and without doubt her cult has had support, fractures, and dissolutions throughout its history. It is far from homogeneous. Whether devotees welcome it or not, Santa Muerte as a popular icon has transcended borders in different ways, one of which is the path of migration. In the city of Phoenix, we find a very widespread cult becoming a common benchmark in almost all the religious shops in the west of the city; for its part, in Los Angeles, California, there at least a dozen cult locations. In addition, in Des Moines, the capital of Iowa, or in New York, we find some altars placed in particular areas where Hispanics practice the cult. In almost all these cases, and in Mexico as well, the cult offers cleansing services, spiritual counseling, loving moorings, and card and tarot readings.

[1]The chapters of Chesnut 2012 are organized by some of these colors: blue for understanding, white for purity, black for protection from negative energy, red for love and passion, gold for money and success, purple for healing, and green for law and justice.

5.2 "Here We All Belong": Migrants and the Cult of Santa Muerte in Guadalajara

The cult of Santa Muerte is not foreign to Mexicans. On the contrary, it has been present in Mexican geography since pre-Hispanic times, bursting with everyday life. It was set up among the living to defend their place, signaling the prelude to a new life, the beginning of the path to reach "the place where the truth lives." In that way, migrants cling to the White Lady in order to guarantee "a new life." "We ask all the saints to help us arrive. Here in Guadalajara, we know that they pray to Santo Toribio and Santa Muerte," assert the Salvadorians who have arrived in Guadalajara on "*The Beast*," the railroad used by Central American migrants and Mexicans to cross the Mexican Republic.

On its long journey through the country, *The Beast's* tracks cross many different cities. The capital of the state of Jalisco is no exception. *The Beast* crosses the Guadalajara Metropolitan Area (GMA) through five towns (Zapopan, Guadalajara, Tlaquepaque, Tlajomulco, and El Salto), touching 48 neighborhoods: 11 from Zapopan, 16 from Guadalajara, 17 from Tlaquepaque, two from Tlajomulco, and two from El Salto. In the metropolitan area, we see various connections between the migrants and the Guadalajaran population. These range from the rejection of "foreigners" and the criminalization, victimization, and discrimination of the migrants; to help for the poor through Non-Governmental Organizations (NGOs); to innumerable cultural and religious practices and narratives that create a multicolor mosaic, of which the cult of Santa Muerte forms a part.

When the migrants enter the GMA, they get off the back of *The Beast* where the streets Juan de la Barrera and Colima cross in the heart of Las Juntas, a municipality of Tlaquepaque. "We know that in this place we must get off, because the police will not arrest us if they catch us. They will [instead] extort us like they did most of the trip," recalled a group of "travelers."

Once down the railroad, the economic exiles walk toward the shelter El Refugio in the hills of Cerro del Cuatro or toward the soup kitchen of the NGO FM4-Paso Libre (Free Pass). At El Refugio, which is located about 15 min on foot from the curve where they get off the train in Las Juntas, the migrants receive food, clothing, and lodging for three nights. At Paso Libre, they access food, clothing, and, if necessary, overnight shelter in the facilities of FM4 located in Calderón de la Barca Street close to Minerva Square.

Regardless of how long they stay in Guadalajara, immediately after they get off of the railroad, many of them proceed to make the sign of the cross at the Las Juntas temple of Santa Muerte, where they are welcomed by Miguel and Sonia, the "owners" of the temple and administrators of the cult. "From Guadalajara, we leave with the blessing of Santa Muerte," says Tomás, minutes before becoming lost in numerous train tracks. Those who leave GMA arrive to "take" the train by the Pacific route, the longest but the "least dangerous" in order to reach the north border of Mexico.

In the GMA, there are three Santa Muerte temples: two located in Tlaquepaque, and one in Tonalá. In this town of pottery (Tonalá), the temple is located in Cerro del Gato; those in Tlaquepaque are in Las Juntas and Las Pintas. The temple in Las Pintas was inaugurated on July 22, 2012. It is by the old road to Chapala at the crossroads of San Onofre in the neighborhood of Huizachera, the place where poverty and crisis live together with devotion and hope. Here the municipal services are conspicuously missing. The dusty streets surround hundreds of half-finished houses. The fetid odors emanating from the rivers that drain from the metropolitan area fill the air. Inside the temple, there are two lines of benches with six rows each. Various posters give life to its white walls. On the wall, next to the old altar hang two crucifixes with Christ on each one. The center is dominated by a large statue of Santa Muerte crowned and covered in white tulle and placed inside a crystal case of glass—in her hands lie the world and a scythe. Next to her is a collection box with the inscription: "Please contribute to the monthly payment for the house of Santa Muerte. Thank you." In the center of the temple there is a baptismal font. At the entrance black, red, and white candles are sold. Before entering the sanctuary and coming across the evil-looking guard, one can read on the side of the entrance: "Santa Muerte spread out your hand and put away your sword. Change pain to joy. Remove dangers and ills from your devotee, who looks for the light to make life longer before departure."

The sanctuary of Las Juntas rises to the side of the route of the railroad. Even though it has been expanded on two occasions, it is still smaller than Las Pintas. Nevertheless, it receives the larger number of parishioners. Above all, the 22nd day of each month is when parishioners crowd around to participate in mass; dozens of the faithful attend the official mass by Priest Ricardo, a plumber who doesn't charge for his services but only receives "help" that the assistants collect, because, as he says, "I am a healing priest for the soul, not for the pocket." Inside the temple various wooden benches, chairs, and images are arranged everywhere. At the back stand two figures of Santa Muerte, like silent witnesses to the ceremony. A small table serves as an improvised altar on which all of the ritual objects are placed. People approach with various bouquets of flowers and some full bottles of water to be blessed, as well as fruits, incense, alcoholic drinks, coins, sweets, and candles. On the wall in the back, two crosses hang. On a pedestal is an image of a Sacred Heart of Jesus statue, and at the feet of Santa Muerte is a candle without a pedestal that complements the "altarpiece" that frames the scene. The faithful carry scapulars and "wear tattoos of the Lady," says a 20-year-old woman, who reveals her tattoo on the upper left of her back. In reality, the mass is very similar to those carried out in the Catholic Church. In it we see the Initial Rites, the Liturgy of the Word, the Eucharistic Liturgy, and the Rite of Farewell. Even if throughout mass the same structure of the Catholic Church is maintained, during the Santa Muerte mass, she is mentioned in the first reading of the sermon and the Gospel; outside of these moments, Christ is the principal element of the ritual.

For Miguel, who is a construction worker and owner of the small house that is home to the temple, the crowd in the facility is the result of the new statue of Santa Muerte, which is nearly three meters high and was brought from Mexico City: "now with this statue the people come much more." There are even days in which two rosaries are offered in the morning and the afternoon. The enormous statue of "The White Girl" is seen in the windowsill of the roof of the house, where they constructed a dome in order to observe it from a distance.

The devotees of Santa Muerte use various books to pray (for example, *The Little Book of Prayers Dedicated to Great Santa Muerte, My Protector*). Inside of them, it is possible to read:

> Thank you, my dear, for permitting me to see the light of a new day, as thanks, I believe in you, I offer you this prayer so my petitions are heard, I trust in you.

> O, Santa Muerte, angel of God, I give you joyful thanks for the favors that you have given me, especially in this day of action and reflection. I pray for my dear loved ones and for my enemies, for the peace of the earth and for the cleansing of my illnesses. Amen.

To get more participation from the parishioners on the days of the rosary, Miguel provides assistants with various photocopies with the prayers that they should recite.

The saints and virgins whom the migrants worship and ask for "a little help" to travel north are numerous: Juan Soldado, whose cult arose in Tijuana from the 1930s, when Juan Castillo Morales, a military subordinate accused of raping and killing a young girl in that city, was led to the Panteon of Puerta Blanca (a cemetery) to appeal to the law for escape; Toribio Romo González, born in 1900 in the shantytown of Santa Ana of Guadalupe, a community in Jalostotitán, Jalisco, killed at the hands of federal forces on February 25, 1928 in Tequila, Jalisco during the Cristero War; Jesús Malverde, the saint of drug trafficking, who was born in Mocorito, Sinaloa in 1870, and who is remembered as a bandit that robbed the rich and gave to the poor; Teresa Urrea, the Saint of Caborca, born in 1873 in Santiago de Ocoroni, Sinaloa; and the Fidencio Boy, who was born in 1989 in Yuriria, Michoacán.

What sets Santa Muerte apart? In the Church of Santa Muerte, there is room for men, women, homosexuals, elderly people, immigrants, poor people, drug addicts, alcoholics, punks, emos, gang members, ex-convicts, drug traffickers, indigenous people, sex workers, street children, and those who are ignored by other churches. "The Godmother" opens her arms to all. She does not judge. Here there are no good or bad people; all who take interest in her protection are given strength at the cult of Santa Muerte. "*Candy* [as they call the new statue] helps and consoles," says Sonia, Miguel's wife. According to estimations of priests of this church, there are two million followers of Santa Muerte in Mexico. The actors that live at the margin of the law have taken possession of the symbolic dimension of the deity: it is not only about popular devotion of socially marginalized sectors of society, nor emerging actors of social exclusion; this church also allows priests to marry, promotes the use of male and female condoms and emergency contraception, supports abortion in cases of rape, and is against the myth of virginity. In the church of Santa Muerte, "all belong."

5.3 Prayers to Death: A Trivium Analysis to a "Santa Muerte" Book of Devotions

5.3.1 Prayers of Resistance

In the following section, I, Eduardo García-Villada conduct a three-pronged analysis (*trivium*) of the sociolinguistic aspects of the Santa Muerte prayers (Joseph 2002),[2] then offer an explanation of the theological aspects of the prayers based on the social relations that the texts bring to the context of popular religion in contemporary urban Mexico. Given that contemporary Mexico is a product of the Spanish conquest, it is logical to maintain that religious expression can be analyzed through a lens that shows the dynamic tension between conqueror and conquered;[3] that is, between the official religion and popular religious manifestations. In the case of Santa Muerte's prayers and rituals, this conflict is played out in private, pagan, insurgent, and delinquent spheres where resistance against the conquest is manifested. I argue that by looking at the practice of the Santa Muerte cult and analyzing the prayers to Santa Muerte, we find continued resistance against the conquest of Mexico that manifests itself as a refusal by a marginalized sector of the population to follow the religious practices of the conqueror.

As a means to understand the issue of power relations in Mexico, where marginalization and resistance occurs, we can look at the disciplines of anthropology and sociology in the works of Scott (1985) and Martín (2017). Based on a discourse analysis of interviews given by Santa Muerte devotees in two documentary films about Santa Muerte, Martín (2017) argues that "Santa Muerte's devotees frequently employ her dark side not because they are coded as suspect and criminal in Mexican society, but rather as a means to acknowledge their representation as marginalized subjects and resist it" (pp. 10–11). In the ethnographic research that led to his book *Weapons of the Weak*, Scott (1985) listed the "everyday forms of peasant resistance" that a farming community in Malaysia employed to resist the external efforts from power elites to industrialize rice production. Attitudes and actions such as "foot dragging, dissimulation, false compliance, pilfering, feigned ignorance, slander, arson, sabotage, and so forth" were all indicators of farmers' discontent. "These … forms of class struggle have certain features in common. They require little or no coordination or planning; they often represent a form of individual self-help; and they typically avoid any direct symbolic confrontation with authority or with elite norms" (p. 29). We see similar actions of resistance from Santa Muerte devotees in the accounts of Perdigón (2008), Roush (2012), Chesnut (2012), and Martín (2017). But unlike Scott's (1985) "ordinary weapons of relatively powerless groups" that are used to resist a hegemonic system in agriculture, Santa Muerte devotees resist

[2] Some books and articles about Santa Muerte include selections or excerpts of prayers (Perdigón 2008; Chesnut 2012), but to date there is no written record of a linguistic analysis of the prayers.

[3] Special thanks to historian and sociologist, Dr. Gustavo Gutiérrez, for his suggestion to include the conqueror-conquered theme.

the hegemonic system of the conquest. I contend that these actions of resistance that Scott (1985) indicated need to be understood from an historical context, where vestiges of history and power struggle are present in the language of Santa Muerte prayers.

In what follows, I attempt to show how prayers to Santa Muerte perform resistance to conquest and conqueror. To do so I employ a threefold analysis of these prayers, utilizing the three lowest divisions of the seven liberal arts—grammar, logic, and rhetoric—which were collectively known as the *Trivium* in medieval universities.[4] Of particular help to this analysis is the use of the Trivium in the reading process suggested by McFarland (2017) with regard to viewing and interpreting films. Accordingly, the *grammar stage* is an understanding of the texts that answers questions about the basic details of the prayers. The *logic stage* is an understanding of the message(s) in the texts that aims to identify the message(s) that the prayers are presenting. Finally, the *rhetoric stage* is the interpretation of the texts that focuses on what we say about the prayers. In summary, my analysis centers on the text of the prayers, the message(s) that they convey, and the interpretation of those texts and messages.

5.3.2 The Grammar Stage

For the *grammar stage* of this analysis, I classify the types of prayers and number of instances in which the pronoun *vos* in Spanish (*you* or *thou* in English) is used in each of the prayers. In certain regions of Latin America there is a linguistic phenomenon called *el voseo*, which refers to the use of the pronoun *vos* to address a singular interlocutor. Although the use of the pronoun *vos* originally had a plural value to refer to a group of people, it is presently used as singular, alternating with two other singular pronouns, *tú* and *usted*, depending on several sociolinguistic variables. Speakers then have three pronoun options to address someone who is of a different social stature, age, or gender. When addressing the saints, God, and the Virgin Mary, it is typical to use *tú* or *vos* to show familiarity or respect through the pronouns that are used.

When we look at the history of the evolution of Latin into Spanish, we find that the second-person-plural pronoun *vōs* from Latin became *vos* in fourth-century Castilian Spanish. Linguistic changes in pronoun usage continued, and by the early thirteenth century there was evidence (as seen in "The Poem of Mio Cid," the oldest Spanish epic poem), that *vos* was being used as a singular pronoun to refer to the nobility, whereas *tú* was being used to address one's family and close acquaintances as well as people of inferior rank. In the first half of the fifteenth century, the form *vuestra mercéd* ("your grace") appeared and subsequently underwent several

[4] "Logic is the art of thinking; grammar, the art of inventing symbols and combining them to express thought; and rhetoric, the art of communicating thought from one mind to another, the adaptation of language to circumstance" (Joseph 2002, p. 3).

Table 5.1 Old and modern uses of vos

Conversation direction and social rank of interlocutors	Old uses	Modern uses
From superior to inferior	*Tú*	*Tú/vos*
From inferior to superior	*Vos*	*Usted*
Among superiors	*Vos*	*Tú/usted*
Among inferiors	*Tú/vos*	*Usted/vos*

changes, after which it morphed into the second-person, formal, singular pronoun *usted*. In eighteenth-century Latin America, the familiar *vos* with singular value alternated with *tú* and *usted*. Table 5.1 illustrates the old and modern uses of *tú*, *vos*, and *usted*, based on the social rank of the speaker and the person being addressed.

Several sociolinguistic factors also play a role in the use of *vos*. Social variables such as age, gender, socioeconomic status, and social rank determine pronoun use in a conversation. Historical and cultural aspects such as indigenousness, nationality, and levels of education do so as well, as does the degree of formality and informality of a situation, depending both on whether the language is colloquial (vernacular), formal (erudite), intimate, or familiar, and also on the distance or closeness among the speakers.

In the history of Latin America, the regions that had longer contact with Spain are more likely to use *vos*. The three regions that predominantly use *vos* in everyday informal communications are the Rio de la Plata region (Argentina, Uruguay, and Paraguay), the Andean region (Colombia, the Andes of Venezuela, and Chile), and the Central American region (Guatemala, Nicaragua, and Costa Rica). Recently in the U.S., some Central Americans have protested that Mexican Spanish is the norm for Spanish spoken in the U.S., where *vos* is decidedly not used. Central Americans who use *vos* want to preserve the form to ensure that *vos* is at least recognized, even if it is not the standard (Alvarenga 2016). It is also important to point out here that the contemporary use of the pronoun *vos* in Latin America comes with its idiosyncratic corresponding verb endings that, although derived from the Old Castilian, are not the same verb forms present in the Santa Muerte prayers.

I center my analysis in the everyday use of language, particularly the intimate act of praying, with a focus on the old forms of Castilian Spanish. The collection of prayers included in this review consist of 38 prayers divided in two sets. The first set is a corpus of 20 prayers collected during two Drake University travel seminars to Guadalajara in 2014 and 2016.[5] The second set of prayers is a corpus of 18 prayers, divided into two groups, that Katia Perdigón (2008) published through the Mexican

[5] Participants in the seminars visited two Santa Muerte temples at "Las Pintas" and "Las Juntas," as well as communities in Tlaquepaque, Jalisco, Mexico. The 19 prayers collected in 2014 were in a spiral-bound booklet of word-processed and photocopied pages printed from a dot matrix printer. The booklet was laid out on a table in "Las Pintas" for public usage, and I was allowed to take digital pictures of each page that were later used for transcription and analysis. In 2016, a set of stapled photocopied pages of a rosary prayer was given to me by the keeper of the "Las Juntas" temple.

National Institute of Anthropology and History (INAH) that are included in an appendix of her book.[6]

The majority of the prayers in the first set use the familiar intimate *tú*, with the exception of three prayers ("Prayer to Protective Spirits," "Divine and Powerful Saint Death," and "Rosary to Holy Death"), which use *vos* with singular values. Of particular interest is the prayer "Divine and Powerful Saint Death," which is written in the form of a jaculatoria, or short and fervent prayer with repetition similar to a psalm or a litany. This prayer uses the archaic singular pronoun *vos* with plural verb endings (*concededme* [*vos*] — "grant me").

> *Luz que dirige su brillo hacia mí, concededme el amor y el respeto al supremo y eterno dios viviente entre nosotros ....*
>
> Light that directs its brilliance towards me, grant me the love and respect to the supreme and eternal living God among us ....

The archaic *vos* verb form *concededme* is used 11 times in all, with the prayer concluding with four instances of the verb form *concédeme,* which is the contemporary *tú* form of the request "grant me."

The second set of prayers contains two prayers that use *vos* with singular values ("Daily prayers to Santa Muerte" in the first group, "Other prayers without title" in the second group), as well as one novena that uses *vos* with singular values ("Novena to Santa Muerte" in the first group). This novena is one of two in the second set, though the only one that uses the *vos* pronoun (as well as the *tú* pronoun). One example of *vos* from this set is *Dadle fin* ("Put an end"):

> *Dadle fin a esta súplica, Muerte Protectora y Bendita por la virtud que Dios te dio ....* (Perdigón 2008, p. 144).
>
> Put an end to this supplication, Protective and Blessed Death for the virtue that God gave you ....

Just like a Roman Catholic novena, this novena is a set of prayers that are prayed on nine consecutive days.

5.3.3 The Logic Stage

The "Logic Stage" of the Trivium analysis focuses on the message(s) that these prayers convey. In the prayers to Santa Muerte, Santa Muerte plays the role of an advocate; that is, she is a lawyer (*abogada* in Spanish). She is also addressed as friend, sister, and mother. However, the intimate or endearing titles that devotees

[6]In the appendix, the author classifies the prayers in two groups: 11 prayers that are prohibited by the Catholic Church and six prayers according to the devotional for the Santa Muerte Traditional Catholic Church, Mexico-US. Unlike the prayers from "Las Pintas" and "Las Juntas" (i.e., the prayers from the first set above), which have numerous spelling errors that are indicative of the low literacy levels of the person(s) who wrote or transcribed them, the prayers from this set are free of grammatical and orthographic errors.

give her in everyday speech (*flaquita*, *niña blanca*, *bonita*, etc.) are rarely present in the text of the prayers.

In general, the prayers follow a formula that goes like this: asking permission from God and the Holy Trinity to address Santa Muerte, then invoking Santa Muerte with the desired prayer or need. The novenas have a different purpose: the first prays for Santa Muerte's help to make a *fulano/a*—an average ordinary person—fall in love with the supplicant, and the second asks Santa Muerte to provide help in living a good and productive life that is free from sin and closer to Jesus Christ.

In the Roman Catholic tradition, praying the rosary includes a set of five mysteries that reflect on the life and miracles of Christ. Each of these mysteries is followed by reciting the "Hail Mary" prayer ten times, which is known as a "decade." In comparison, the five mysteries of the Santa Muerte rosary ("Rosary to Holy Death" in the first set) reflect on the following topics: (1) faithfulness of her chosen devotees, (2) a promise to respect and defend her, (3) a promise not to hide the faith that is professed to her, (4) gratefulness for all the favors and blessings from her, and (5) gratefulness for the health, work, love, and strength that she brings. The text of this rosary is followed by a series of litanies that exalt Santa Muerte as the protector of the poor, of single mothers, orphans, the elderly, of battered women, abandoned children, prostitutes, people of the third sex, the disabled, the kidnapped and disappeared, those who committed suicide, and the dead, among other people. The rosary is doubly important and jarring in the Santa Muerte tradition of Mexico City: first because rosary prayers have always been dedicated to the Virgin Mary and the life of Christ, and second because the praying of the Santa Muerte rosary has become an event that is attended by thousands of people (Roush 2012). In other words, Santa Muerte has co-opted the rosary and displaced both the Virgin Mary and Christ from their own prayer.

5.3.4 The Rhetoric Stage

The final stage of the *Trivium* is the "Rhetoric Stage," which is concerned with the interpretation of these prayers. Since language reflects power relations, I argue that the similarities between Western and Mexican indigenous religious traditions regarding how the texts were produced for and consumed by the elites reflect uneven power relations and also illustrate how anachronistic linguistic forms not only have been preserved but also still prevail in the present-day prayers to Santa Muerte. The ordinary people who wrote the prayers found in Las Pintas near the metropolitan area of Guadalajara are employing linguistic forms that can be traced back to the language that was introduced by the conqueror during the time of the Mexican conquest. They are using the archaic language of old Roman Catholic prayers as a way of "elevating" Santa Muerte prayers, making these prayers sound far more traditional and orthodox than they are, meanwhile swapping out Mary and Jesus for Santa Muerte, and linguistically lifting Santa Muerte above them as superior.

As with Roman Catholic religion, Santa Muerte practice includes a hierarchy of power that situates humans in relation to God and the divine. According to Perdigón (2008), Santa Muerte devotees practice their beliefs in accordance with a hierarchy of divine beings. In that hierarchy, Santa Muerte is above the Catholic saints and the Virgin Mary, who are not necessary to have access to God through Jesus Christ. This hierarchy is used to establish the power relations between humans and Santa Muerte. Given that Santa Muerte possesses such a high position of power, it would be expected that humans would use formal language and the pronoun *vos* when addressing Santa Muerte through prayers, at least if the prayers follow the pattern used in old Castilian Spanish. However, changes in the use of pronouns to address the saints and the divine have switched from the formal *vos* to the familiar/personal *tú*. The presence of *vos* in Santa Muerte prayers found in this study, then, conform to residual or archaic/antiquated (conqueror) Castilian Spanish. Santa Muerte may be seen as an equal, or as divine.

Prayers to Santa Muerte are also highly personal. They convey a sense of dialogue between the individual devotee and the Saint, between the individual and a higher power. The use of diminutives, such as *Blanquita* ("Little White Girl") and *Flaquita* ("Little Skinny Girl"), which devotees use to describe Santa Muerte in everyday life, shows a sign of endearment or having familiarity with her. An analysis of the plural first person subject or object pronouns in Spanish (*nosotros*, *nuestro* = "we," "us," "ours" in English) in the two sets of prayers collected at Las Pintas and Las Juntas in 2014 and 2016 and in Perdigón (2008) indicates that the majority of the prayers are from the point of view of an individual. Only the rosary, the novena, and the prayers from parents to protect their family or missing children are prayers expressed from the collective "we" or "us."

5.3.5 Theological Aspects of Santa Muerte Prayers

The cult of Santa Muerte has adhered to its pre- and post-conquest religious roots. With regard to the latter, its liturgy follows the worship observed in the Roman Catholic Church, including prayers that are mostly petitions for protection and intercession at the private individual level, as well as novenas and rosaries at the public and collective level. With regard to the former, Santa Muerte as saint is elevated to a position close to God and Jesus Christ (Perdigón 2008) that is reminiscent of the status that Mictlancihuatl (Lady Death) holds in protecting the ancestral remains of the Aztec underworld and also in possessing the powers of both creation and destruction. Prayers to Santa Muerte, therefore, simulate the patterns and rituals of Roman Catholicism, while evoking the indigenous ancient songs to death of the early inhabitants of Mexico.

I have provided sociolinguistic evidence to advance the argument that, in the Santa Muerte cult and the language of some of the prayers used for personal and collective devotions of people in need, Santa Muerte is elevated to a superior and divine status that devotees invoke from their human and lowly status. Santa Muerte as a saint and as a deity has the power to destroy and create that is reminiscent of other goddesses in the pantheon of supreme beings of pre- and post-conquest Mexico.

I argue that the language of Santa Muerte prayers, as analyzed in this paper, takes a conciliatory stance on the resistance against religious and economic conquest. This position is taken within the order of power—it accommodates; it is a strategy of survival; it adopts a neutrality; and it assumes tolerance. In the Santa Muerte cult, as in the time of death, everybody is equal and welcome. The very fact that devotees create an independent space for expressing their religious beliefs is an act of resistance. Not resisting is disappearing.

At the conclusion of this analysis of Santa Muerte prayers, there are many unanswered questions. Why do people who pray to Santa Muerte use the antiquated language that is not part of the standard every-day language in Mexico? Why were the beloved Mexican Virgin of Guadalupe and the various saints of the Roman Catholic church no longer sufficient to intercede on behalf of marginalized people, such that they needed to create their own saint? Why do people need a dead entity to protect them from risky living? Is there a conscious knowledge of the history among devotees of the pre-conquest religion to which people want to go back? What prompted the cult to Santa Muerte? Although the answers to these questions are largely unknown or speculative, what we do know is that Santa Muerte is out in the open. A frightening saint in the middle of a Mexico City street, as in my initial sighting of Santa Muerte, and as present and visible as it is in Las Juntas, Guadalajara, and beyond the Mexican borders. The Santa Muerte practitioners are indeed following the words of the rosary reflections not to hide their devotion.

5.4 Saint Death and the Medicalization of Death

Our two lectures on Santa Muerte were among the most engaging and informative in our 2015–2017 lecture and dialogue series. Nevertheless, they left us with one essential question: What does the cult of Santa Muerte have to do with the medicalization of death? Although we knew that it was important to consider the cult of "Saint Death" within the context of a lecture and dialogue series on death and dying, particularly one that focused on recent change, we did not know whether and how to connect it to bioethics.

It was about halfway through the series, when it occurred to us that if the medicalization of death is a way to exert greater control over death—to have more say about when and how one dies—then the cult of Santa Muerte is similar in end. Praying to "Saint Death" is, among other things, a means of protecting oneself from death and other forms of suffering and misfortune. It is not surprising, then, that death is the saint to which many marginalized sectors of Central American society have turned.

We are not scholars of Mexican society. But it seems to us that for many of the devotees of Santa Muerte in the Barrio Tepito, medicalized death is neither a possibility nor a concern. For these followers of "Saint Death," both traditional religion and the central state have failed them. In a time when acute insecurity and fear of death are constantly present, devotees of Santa Muerte have turned to her as a saint, if not deity, to protect them against death. She protects them and provides for them when no one else can or does. This too is a religious response to modernity with regard to death and dying, if not a response, in a sense, to the medicalization of death. It is an attempt to control death, just like bioethics.

References

Alvarenga, Daniel. 2016. Vos vs. Tú: 4 Central Americans on proudly reclaiming Voseo in the United States. http://remezcla.com/lists/culture/central-americans-reclaiming-vos-california/. Accessed 22 May 2016.

Chesnut, Andrew R. 2012. *Devoted to death: Santa Muerte, the skeleton saint*. Oxford: Oxford University Press.

Joseph, Miriam. 2002. *The trivium: The liberal arts of logic, grammar, and rhetoric: Understanding the nature and function of language*. Philadelphia: Paul Dry Books.

Kroger, Joseph, and Patrizia Granziera. 2012. *Aztec goddesses and Christian madonnas: Images of the divine feminine in Mexico*. New York: Routledge.

León, Juan Luis. 2007. *La muerte y su imaginario en la historia de las religiones*. Bilbao: Universidad de Deusto.

León-Portilla, Miguel. 1993. Those made worthy of divine sacrifice: The faith of ancient Mexico. In *South and Meso-American native spirituality: From the cult of the feathered serpent to the theology of liberation*, ed. G. Gossen, 41–64. New York: Crossroad.

Martín, Desirée. 2017. Santísima Muerte, vístete de negro, Santísima Muerte, vístete de blanco: La Santa Muerte's illegal marginalizations. *Religions* 8 (3): 36. http://www.mdpi.com/2077-1444/8/3/36. Accessed 24 May 2017.

McFarland, Mitzi. 2017. Viewing film for the purpose of interpreting it. http://www.westga.edu/~mmcfar/viewing_film_for_the_purpose_of.htm. Accessed 24 May 2017.

Pennock, Caroline. 2012. Mass murder or religious homicide? Rethinking human sacrifice and interpersonal violence in Aztec society. *Historical Social Research* 37 (3): 276–302.

Perdigón, Katia. 2008. *La Santa Muerte, protectora de los hombres*. México: Instituto Nacional de Antropología e Historia.

Roush, Laura. 2012. La informalidad, la Santa Muerte y el infortunio legal en la Ciudad de México. In *Informalidad urbana e incertidumbre: ¿cómo estudiar la informalización en las metrópolis?* ed. F. Alba, F. Lesemann, and L.C. Bustamante, 221–245. Mexico: Universidad Nacional Autónoma de México, Coordinación de Humanidades, Programa Universitario de Estudios sobre la Ciudad.

Scott, James. 1985. *Weapons of the weak: Everyday forms of peasant resistance*. New Haven: Yale University Press.

Part II
Medicalization and Religion

Chapter 6
Christians Encounter Death: The Tradition's Ambivalent Legacies

Lucy Bregman

Abstract The central focus for Christians has been on the death of Jesus Christ; it is his dying, death, and resurrection that has shaped what Christians have believed, taught, and hoped. Some implications and limits of this model for death include that it is difficult to consider death a neutral, "natural" event, given that the Christian focus has been on the violence, destruction, and link to sin involved in Jesus' death. Although diversity in the practice and experience of Christians is recognized, the central issues regarding Jesus' death persist within the Christian tradition's understanding of how death fits within the totality of human existence.

6.1 A Tradition and Its Focus: The Death of Jesus

Years ago, I received in the mail a glossy advertisement for a Christian book club. There were hundreds of possible selections in it. Lots dealt with family life—parenting, etc.— and many with obscure passages in the Bible. Most were meant to show how Christian faith could provide resources and strength for life's difficult situations. Indeed, there seemed to be no issue or situation in human life missing from the book club's brochure. Except: none of their books dealt with death. None. It was as if the answers were so obvious that Christians did not need books on death.

This was at a time when there were plenty of Buddhist books on conscious dying, so I was troubled that there were no corresponding books to guide dying Christians and their companions. Maybe there was an assumption that Christians are members of congregations, communities over time that can support and console their members, while Buddhists are more solitary and therefore must rely on books. Yet at almost all times in the past, Western Christians have had relevant resources to consult and appropriate, an entire tradition of *Ars Moriendi* written to help the dying face their future. The twenty-first-century book club illustrates how distant this

L. Bregman (✉)
Religion Department, Temple University, Philadelphia, PA, USA
e-mail: bregman@temple.edu

T. D. Knepper et al. (eds.), *Death and Dying*, Comparative Philosophy of Religion 2, https://doi.org/10.1007/978-3-030-19300-3_6

tradition has become, for the task of actively transferring it into a resource for ordinary people has seemingly become just too daunting.

Being faithful to a tradition that has had plenty to say about death (not all of it helpful), while saying something that makes sense today, is still a work in process. What was once said may still be relevant, or it may prove that not all of the past can be turned into a helpful resource for us. This is the meaning of the "ambivalent legacies" in my title. There are beliefs and devotional practices, once very popular, that would make even the stodgiest, most old-fashioned Christian cringe today, while other ways to encounter death as a Christian are worth attempting to understand and appropriate.

This essay attempts to identify and interpret the core themes in Christian understandings of dying and death, themes that contrast sharply with today's medicalized vision, as well as with the cheery focus on Christian families and parenting, health and fulfillment, that dominated that brochure. Moreover, these themes—theological, liturgical, spiritual—are not necessarily what the average Christian, down through the centuries and in many different lands, considered most important to his or her own faith. So, this is not an ethnographic study of "lived religion," but an analysis of an intrinsic inner core that has been acknowledged as such within the tradition.

"Whenever you do this, you proclaim the Lord's death until he comes," wrote Paul (I Cor 11:26). It is still the death of Christ around which Christian reflection and practice about death centers. If this is an essentialist approach, it is because on this matter the consensus of Christians has itself been to acknowledge an essence, expressed in creeds, catechisms, and official doctrinal teachings. Documents such as the Apostles and Nicene Creeds incorporated the narrative of Jesus' death, burial, and resurrection at their center. Although, as we will see, there is some flexibility as to how a full-blown theology of death might look, that is not to say that there are no core, central ideas considered normative for all. This essay will consider especially how, during the last century, challenges to some long-standing ways to proclaim and imagine Christ's death reared up, and how thoughtful Christians faced them. This is a process that is ongoing at the levels of theology, pastoral care, and liturgical practice. In some cases, the emphasis has been on reducing an intense exclusive focus on dying, such as when, in the wake of Vatican II, the Roman Catholic Church renamed the sacrament of Extreme Unction as "Annointing of the Sick," a ritual focused on healing. In other cases, as we will see, Western Christians have been particularly concerned to deliteralize portrayals of the afterlife, and also on de-Platonizing Christian views of persons, so as to affirm a more holistic understanding of who we are as created, embodied beings.

It helps also to acknowledge the contemporary medicalized vision of death as a starting point utterly discontinuous from this tradition, or indeed from any traditional understandings around the world. Medicalized death refers to the dominance not just of doctors and hospitals, but of an understanding that equates death with the biological ending of life, and therefore as an entirely organic event. Frequently, this is combined with a different image, that of a battle against death waged by scientists/doctors who hope that death, like smallpox, can be defeated and humans can live death-free lives. When we hear of an important person who "lost his battle with cancer," it is this imagery that places our sense of death's meaning in a militarized, medicalized context.

Christians supported medicine and cared for the sick, as monasteries welcomed vulnerable persons of all kinds and Christian missionaries founded hospitals in many places of the world. Visitors to Beaune in France can tour the hospital, founded by lay persons in the 1400s, which functioned as a hospital right up until 1971. Christians never ceased to pray for healing (Gusmer 1984) as well as work toward it. There is no absolute opposition between "religion" and "science" here. But to think of death as a medical problem, a condition that might—with enough skill and technology—be fixed, is to think in a way no one, anywhere, had ever imagined until the twentieth century in industrialized countries. The patients of Beaune could visit its chapel and contemplate Van der Weyden's wonderful painting of the Last Judgment, to remind them of the ultimate meaning of their life and death. By contrast, our medicalized understanding forms a kind of barrier, preventing us from seeing and apprehending whatever previous cultures, religions, and traditions knew about death. It has made it very difficult to see and learn from them, even when many of us are convinced our current society desparately needs wisdom rather than just new technology.

A thoughtful book by Allen Verhey, *The Christian Art of Dying: Learning from Jesus*, starts from this claim and represents the tradition, its contributions, and its failures in the light of what I also argue is the correct starting point (Verhey 2011). The relative paucity of works such as this is an unfortunate sign of what that book club brochure illustrated so vividly, namely, the gap between past focus and present preoccupations. Medicalized death is itself the subject for other essays in this volume, especially those focused on bioethics. But we will try to pierce the barrier, to discover what lies beyond medicalized death that might or might not be helpful for persons today. Admittedly, this means helpful for those persons fortunate enough to live where hi-tech medicine and abundant pharmacological resources are available and taken for granted. As with most of the other essays in this volume dealing with medicalized death and bioethics, we write for the "developed" world and its inhabitants.

The most important human death for Christians is the death of Jesus. This has shaped what Christians want to think and teach about death, although it is acknowledged that his death is unique. There are a few very basic things about how unusual Jesus' death was, which those of us steeped in contemporary issues should keep in mind. First, the unnatural quality of his death: like the death of Socrates, it was a kind of judicial murder, and the medical causes were of no interest to any who recorded the event. There was not even a doctor present to certify his death, just a Roman soldier to stab his corpse. Second, what followed Jesus' death was not a funeral, nor mourning. While dying, death, and bereavement seem to be the cluster of topics discussed today, what follows death for Christians is resurrection, based on the witness of the first followers of Jesus. Contemporary interest in Holy Saturday, the day between death and resurrection, Good Friday and Easter, tries to include some attention to grieving; this is a recent attempt to respond to what now seems a missing element in Christian spirituality (Lewis 2001). It is Jesus' trajectory of death, descent, and resurrection that mattered in liturgies of baptism and in creeds, not what others were doing or feeling. While Roman Catholic theologian Karl Rahner held that the paradigmatic Christian death is that of a martyr, this in turn relies on a derivative model (Rahner 1961, pp. 89ff.). The martyrs themselves followed their Lord and Savior into death. From this starting point, specific to Christianity, there are some basic consequences and later developments that follow.

1. Doctrines and rituals focus first and foremost on Jesus' death and resurrection, not his birth, his teaching, or his activities. What happened to Jesus will happen to the Christian linked to him, identifying with him—in baptism, through the Lord's Supper, and in preparation for one's own death (Schnackenberg 1964; Paxton 1990). Although it could not have been just anyone's death and resurrection, the sources we have of his life are shaped through and through by the shadow of Jesus' impending death. The authors of the Gospels knew the ending and the entire plot when they wrote the earlier chapters. D.H. Lawrence's *The Man Who Died* (1929) follows this gospel plot; at least his Jesus dies (and finds new life and sexual fulfillment with an Isis priestess!). Jesus' death here signified the repudiation of the West's ascetic ethic, bourgeois morality, and theistic worldview—but Lawrence could recognize a death as central to the narrative of his protagonist.
2. Because of this intense use of Jesus' death as a model for what human death means, it is hard to say that for Christianity "death is a natural event." It was certainly accepted as a part of life, and if that is what we mean by "natural event," then there is no problem. Indeed, by contemporary standards, it was over-accepted, wallowed in to the neglect of other aspects of Jesus' life and ministry. But the category of nature—or maybe Nature with a capital N—is just not where this tradition started, and it does not illuminate what Christians have said. When a seminary student told me, "What else is there to say about death except that it is a natural event?" he was certainly wrong historically and probably pastorally. Whatever the role of Natural Law theory elsewhere, Christians have not been focused on "natural" when it comes to death. In the famous "Dies Irae" poem by Thomas of Celano, at the Last Day, the Day of Wrath, "Death and nature will marvel," subject to the same ultimate Divine Decree as are we (Celano n.d.).
3. By and large, the philosophical framework used to elaborate doctrinal statements—those believed, confessed, and taught by the Church—has been taken from Plato, with modifications. This means that dichotomies between material and transcendent realms, and between soul and body, have dominated the language of formal thinking. Whether we like it or not, whether we consider this language less than "authentically Biblical," this is an historical reality. However, it was never simple Plato; death and resurrection really were embodied experiences, and Incarnation itself as a category clearly breaches conventional Hellenistic dualisms (Wolfson 1965). As Jaroslav Pelikan showed in his lectures on "The Shape of Death," there was no one way to imagine death's meanings theologically in the ancient church, nor any one way to include Plato's categories into Christian teaching (Pelikan 1961).

4. Add to these Plato-based categories the dichotomy between this age and the world to come. The sense of a future toward which God's salvation is headed, and which potentially relativizes all the current structures and preoccupations of the world, is part of a theological framework that includes personal death but goes beyond it. Remember the "Dies Irae" is not just about my own death, although it was part of Roman Catholic funeral liturgies for centuries. It is about the end of everything other than God. The words of an 1899 preacher at a funeral—"Last week he was in his office. Today we bury him. Are you living under the power of the world to come?" (Ketcham 1899, p. 18)—were probably spoken at funerals 1000 years earlier (see Bregman 2011 for more examples similar to this one). Whether this world to come is a cosmic end time for all things created, or whether it is focused on one's own personal immediate transition to an eternal realm, the intended result is to make the office look unimportant by comparison. However, a large part of the shift in traditional Christianity's way to interpret death as transition has been to de-emphasize "the world to come" as the sole goal or focus for Christian living. If there is a redeeming value to that Christian book club brochure, it is that it gave this life (of families and parenting, etc.) a value in the light of God's ultimate purposes, without setting up an opposition such as the preacher assumed, or that D.H. Lawrence built his story around, with the Isis priestess and her love as the earthly substitute for the traditional world to come.

6.2 Implications and Some Traditional Issues

There was always leeway given to thinking on the topic of death. There is no one Christian theology of death, according to Jaroslav Peilikan (Pelikan 1961). There were a variety of imageries, allegories, and analogies open to religious thinkers. Some were more enthusiastically Platonist than others, or more collective-communal rather than individualistic. There were a few options that were ruled out, by means of official pronouncements. Origen, the third-century Christian intellectual, apparently advocated some form of reincarnation, or, more likely, multiple successive existences in various spiritual realms or levels rather than successive embodied lifetimes here. This was for the purpose of purifying the soul, so that it could reach eventual perfection. Such a view was eventually condemned, and there is not reliable evidence that "Christian reincarnation" was ever an established belief anywhere. When the Protestant Reformation ruled out Purgatory as a non-Biblical and erroneous teaching, there was still, eventually, room for the hope that God will purify and perfect the souls of those whose earthly lives were flawed and incomplete. How this would look was left vague. Unfortunately—as many contemporary Christians regret—the Church affirmed again and again that Hell was eternal, that universal salvation was not a permitted Christian hope, and that the final division of humanity into sheep and goats, saved and damned, was not to be abrogated (Walker 1964).

Yet these debates remained at an official level that may not have reached deeply into how people encountered death. In spite of these theological emphases, there is massive evidence that throughout the ages most Christians experienced what historian Phillipe Ariès called "tamed death." They died as if moving into the next room, without trauma or terror or disruption. Ariès believes that while tamed death was the norm of the past, "today it has become wild," that we have made it uncontrollable and overwhelming (Ariès 1974, p. 14). Christians wished for a time to prepare for their own deaths, they wished to be buried in or near a church, and to rest in peace until the day of resurrection and judgment, Paul's Day of Christ. Yes, there were times when these expectations were swamped by more frightening scenarios. In times of plague, when sudden death and social chaos interrupted normal preparation rituals, the sense of life's fragility and divine wrath overwhelmed people. Death's wildness triumphed, although not in the way it does for us. We find grim evidence of this in preaching by wandering monks that dramatized the need for dramatic self-abnegation and repentence to avoid the torments of hell. We find this same preoccupation with mortality and evil in stone portraits of the dead as corpses or skeletons, rather than laid out peacefully in their armor or finery. Long, long after, it is impossible to guess if the bishop, who had himself carved as a skeleton for his gravestone in the cathedral, was more pious than other churchmen who did not. Maybe he had been an s.o.b. in life, and just wanted to make all the survivors miserable too as they contemplated his memorial.

This obsessive focus on death's destructive power, associated with the European late-medieval period, raises another question that has renewed relevance in the modern era. Although there is language in the New Testament that links death with sin and evil spiritual forces, how far should Christians go in demonizing death, in making death God's enemy, a Satanic force? Yes, a force defeated by Jesus through his resurrection, but with the triumph of the resurrection often overshadowed by the vivid depictions of Death (personified) as destroyer. In the days of tamed death, this may not have had any personal application, whatever its appeal today. Once again, the twentieth century brought re-thinking, although here it was the horrors of twentieth-century history that renewed this theme of death's horror and evil. The narrative of *Christus Victor*, as the title of an influential book by Gustaf Aulén (1967) proclaimed, was rediscovered from ancient sources, and now advocated as the authentic understanding of atonement and of death. Death and Satan, conjoined, were losers in a cosmic battle—a view vividly captured in the almost-final scene of Mel Gibson's film "The Passion of the Christ" (Gibson and Fitzgerald 2004). We look down from God's eye through the earth to Satan below, screaming and snarling in defeat.

An alternative view, more commonly held and preached even into the first half of the same century that discovered the power of Christus Victor, is that Christians should befriend death, welcome it as liberation from the dangers of sin, or from bodily frailty and pain. This is the view that Verhey castigates in his recent analysis of the tradition (Verhey 2011, ch. 7). Should Christians joyfully anticipate their transition to Heaven, as in the early Methodists' model of happy dying? John Wesley's account of the dying of Jane Cooper is used by him to demonstrate that

Christians can reach perfect sanctification in this life, and it serves as an example for others to follow (Wesley 1981, pp. 342–347). Ms. Cooper may have had what today we call a Near-Death Experience, in which the transcendent realm of eternity and Christ eclipsed what was left of her painful bodily existence here. She certainly welcomed death, and the pain of her dying became irrelevant to her. But was this ever truly a norm for everyone? Or is there a space, even within the context of faith, for fear, sorrow, and regret? Once again, a Christian book club that denies dying and death any space at all, loses the opportunity to raise this theological and pastoral concern.

To some extent, the complexity and ambiguity of these possibilities can be traced back to the model of Jesus and his own dying as portrayed in the four canonical Gospels. Jesus is deeply distressed and anxious in facing his own death, as he prays alone in the garden of Gethsemane. His pain continues right up to the moment of his death, according to Mark and Matthew. They do not have him "befriend death," let alone die joyously. On the other hand, they do not introduce Satan as a character in these scenes of the narrative. It is we—or at least Mel Gibson following Aulén—for whom death is wild, who add Satan as a visible presence, to make sure we add the effects of a supernatural horror film to the evilness of death. Yet the narratives of Luke and John show Jesus dying serenely, with a sense of fulfillment in regard to his mission, if not with Jane Cooper's exuberance. In both, he also continues his service to others, consolation for his mother and for the criminal dying next to him (Bregman 1999, p. 74ff.).

We can see these dual stances persisting within the tradition. In the Puritan classic John Bunyan's *Pilgrim's Progress*, the hero of Part I comes to the edge of the river that allegorically separates this life from the next, from his goal of the Celestial City. Stepping into the river, Christian the pilgrim nearly drowns in terror, and yet somehow manages to cross it in faith (Bunyan 1964, p. 143). He is welcomed into the Celestial City on the other side with great honors, and no one scolds him for his moments of faintness and cowardice. Bunyan's Part II was meant to stress the communal nature of Christian life, with a group of pilgrims who travel the same route together, and do not make as many mistakes as the solo hiker. When Mr. Great-Heart, their ideal pastor, prepares to enter the same river, he declares with confidence and hope:

> This river has been a terror to many; yea, the thoughts of it have also often frightened me. But now, methinks, I stand easy…the thoughts of what I am going to, and of the conduct that awaits me on the other side, doth lie as a glowing coal at my heart. I see myself now at the end of my journey; my toilsome days are ended. (Bunyan 1964, p. 282)

I think Bunyan's point is that both experiences are Christian experiences of impending death; neither is condemned. Nor is the river itself personified, made into a demon or a welcoming consoling presence. Bunyan would agree with that early twentieth-century preacher that the nearness of death entirely relativizes the reality of the workplace and even family life. The world to come takes priority. In none of these cases is the grieving of relatives or the medical cause of death of the slightest interest to the writers. Nor is Bunyan describing Ariès' tamed death; death and dying are not the same as slipping out of one room and into another. Mr. Great-Heart is not dying "naturally"; he is dying with Christian hope and trust, which overcomes fear and feels "a glowing coal at my heart." There is also no sign, based on Ariès kind of research, that most Christians worried about hell as their personal destination. However Bunyan personally may have done so, yet the doubts and fears of Bunyan's Pilgrim do not really mirror the popular horrific visions of tortures and demons. Nevertheless, Bunyan's anti-hero Ignorance ends up in Hell, which for Bunyan was an absolutely necessary part of the allegorical landscape. Ignorance, ironically, had a smooth ride across the river, carried by a ferry that supported his denial of his real condition (Bunyan 1964, p. 149).

In short, the big story for Christians is the death and resurrection of Jesus, and we are to fit ourselves within this frame as the little stories (Krieg 1984). There are of course some modern, or at least Enlightenment-era dying individuals, who found this too constraining and abandoned it. Not as Lawrence tried to do, but by neglecting it altogether. David Hume's deathbed meditations are an example, used recently by neurologist-author Oliver Sacks to express his own reflections on his impending death (Sacks 2015). But far more central was the spirituality of the "Passion Chorale," a famous hymn that begins with a depiction of Christ's dying, "O sacred head now wounded," but ends with this personal appropriation:

> Be Thou my consolation
> My shield when I must die.
> Remind me of Thy passion
> When my last hour draws nigh.
> Mine eyes shall then behold Thee,
> Upon Thy cross shall dwell.
> My heart by faith enfolds Thee.
> Who dieth thus dies well. (Gerhardt 1656)

Far from leading to a one-size-fits-all approach, Christians seem to have used this model in ways that could give voice to their own personal struggles and fears, without being completely overwhelmed by these, and with the promise that "Who dieth thus dies well."

6.3 Platonism Challenged: The Dualism vs. Holism Debate

Now let us focus on some of the limits, the problematic aspects of this tradition, especially as contemporary Christians try to turn to it as a guide or resource for thinking better about dying and death today. The first, and most extensively debated issue, is the already-mentioned holism vs. dualism question, and the overall rejection of simplistic Platonism's categories. Traditionally, death is the separation of soul and body. Do we really want to cling to a sharp body/soul contrast, when modern biology insists that we are unified beings, incarnate and also part of a thoroughly incarnated world? Does not the idea of a soul, floating in its own transcendent ethereal realm, betray this deep knowledge? Has not this dualist language been used repeatedly to cast contempt on creation, on the goodness of the world that God created and believes worth redemption? The answer to all these rhetorical questions is: Of course, this has sometimes been the result of too much reliance on Plato, but need this be always the case? It is easy to find even twentieth-century examples of this, such as the position of Christian philosopher Seth Pringle-Pattison in the Gifford Lectures of 1920: "Physical death ought to appear no more than an incident in life…It should bring with it no depressing suggestion of finality…. 'Unbelief in death,' it has been said, 'seems to be the necessary characteristic or concomitant of true spiritual life'" (Pringle-Pattison 1922, p. 208). If the truly spiritual life leads to "unbelief in death" in this manner, then the death of Jesus is a "non-event," and the resurrection is not a miracle but a regular, universal expectation, if its "physical" aspect is overlooked. For many pastors and theologians, a view such as Seth Pringle-Pattison espoused now appears sub-Christian rather than ultra-spiritual.

In the twentieth and twenty-first centuries, the Bible has been invoked to support just the opposite of the Platonic-dualist position most often enshrined by the traditional language. I have tried to track the first attacks against Plato issued in the name of "authentic Biblical views of man," and it seems that by the 1950s the language of Seth Pringle-Pattison appeared to theologians and Biblical scholars as "sub-Christian," ignorant of the deeper appreciation of the goodness and full reality of the material world God created, sustains, and will redeem. A very famous example of such an attack on traditional Platonism-disguised-as-Gospel was the essay, given as a lecture in 1955, Oscar Cullmann's "Immortality of the Soul or Resurrection of the Dead." Here, Jesus' dying and death are directly contrasted to the case of Socrates, who views death as Seth Pringle-Pattison did 2400 years later. For Cullmann, the Bible—both Testaments—proclaims not just holist views of persons, but a strikingly terrifying vision of death. Death is "God's enemy," and fills even Jesus with both "horror" and "terror" (Cullmann 1965, pp. 14ff.). Cullmann lambasts the Platonism that makes almost all Christians squeamish in facing this spectre, and therefore in denial of the truly miraculous victory of God over death accomplished in Jesus' resurrection.

Cullmann's influential essay percolated into many Christians' presentations of theological anthropology and therefore understandings of death. A particularly readable one was David Myers' *The Human Puzzle*, which attemps to tie an "authentically Biblical view of persons" with current psychological theory. Myers may have been skeptical about some of psychology's inflated claims, as well as about residual Platonism, but he devoutly followed Cullmann in proclaiming holism as *the* authentic view of persons (Myers 1978, pp. 73–90). As John Cooper noted, by the latter decades of the twentieth century, this advocacy of an authentic Biblical holism had become heavily entangled with other ideological and political debates. Cooper, summarizing the voluminous literature on this issue, enters a plea for a separation of the historical question (which view did the Biblical authors hold?) from how people now use this language of holism vs. soul/body dualism (Cooper 1989). I find it especially paradoxical that some of the loudest anti-Plato and anti-Descartes voices, strong advocates of a holist view of persons, also espouse meditation practices that themselves depend upon the Indian-Sanskrit version of dualism, the power of mind or consciousness over the merely material. Although we have much more scientific information about the brain than we did when Cullmann wrote, I do not believe this alone is going to solve the essentially philosophical-theological debate for us. At present, it appears that we need some form of dualistic language to describe human experiences of inwardness, and no current research on the brain is likely to abolish reliance on this. This, however, falls very far short of endorsing the classic dualism which treated death as a non-event, with no depressing reference to finality.

Let me clarify this statement. In spite of all the passionate debate over this issue, I believe the problem isn't dualistic language. It's that too often, the soul is invoked—or mind or consciousness—as floating high above the realm of change, suffering, death, and decay, just as in Seth Pringle-Pattison's equation of this with "true spirituality." Yet maybe that's what such language can be good for. At its best, it depicts the condition of the frail 89-year old in the nursing home who is still "sharp as a tack," surrounded by those whose decay on all fronts is so evident. Her mind or soul may soar, gleam, taking flight over space and time, while her body lies in bed almost devoid of movement. Or, taking the more dismal possibility, the condition of the Alzheimer's patient whose physical being is still full of vigor but whose memory is gone, and whom others sense as the empty shell of a beloved person. I'm not saying Plato devised this language of soul and body to account for such experiences, but only that some form of dualism really can do what a consistently holist approach cannot. Stories such as those retold in Maggie Callanan and Patricia Kelley's *Final Gifts: Understanding the Special Awareness, Needs and Communications of the Dying* rely on at least a modest dualism in order to make sense. Here, the dying do seem detatched from their physical surroundings, interacting with figures from the past, dead family members, and "held back" by the pull of embodiment in this world (Callanan and Kelley 1992, chs. 12 and 15). Callanan and Kelley's dying hospice patients join Bunyan's protagonists in awareness that they are about to cross a river or other barrier, and that this is a real experience rather than a non-event for the soul.

6.4 Death as the Punishment for Sin

There are other problems with the tradition's language, imagery, and central ideas that are impossible to avoid. The connection between death and sin is one of them. According to Bonnie Miller-McLemore, once Christians stopped saying this, they lost any distinctive ethical perspective to the whole understanding of death (Miller-McLemore 1988, p. 134). Jesus died and took on our sins, Jesus bore the punishment that should have fallen on us, and Jesus' death is a payment or a propitiation for sin. For "before God we are always in the wrong," and on the Day of Wrath this will be made plain for all. This language is just there in the New Testament; it is already in place and makes death a feared reality even without all the horror and terror language Cullmann heaped upon it. Literalistically, this death-sin link led to speculations that Adam, had he not sinned, would be immortal, death-free. Moving away from this formulation, there nevertheless remains a strong sense that death taints the human condition to the core. Modern thinkers (summarized in a fine discussion in Urban 1995, ch. 5) insist that this is not about particular sins but about existential lack, failure, or misdirected will. Something within us is just basically wrong, and this wrongness ties in to death, or at least our anxiety in the face of death.

The language used for this, however, has been what has been called "juridical," relying on categories drawn from law courts and the criminal justice system. To put this most starkly, we must ask: is death a punishment for sin? I have learned that even to raise this question is to risk others' assumption that I accept this idea in this form.

Curiously the power of this juridical imagery was expressed most forcefully by non-Christians, just at the point in the last century when Christian theologians tried to find alternatives. Two famous pieces of literature by non-Christians demonstrated this: Franz Kafka's *The Trial* (2009 [1925]), and Albert Camus' *The Fall* (1956). Both novels depict guilt, but not over particular, named faults. They lack a God before whom we are in the wrong. Nevertheless, Kafka's K. is to be put on trial, and the nameless narrator of *The Fall* loses his false innocence as he hears a cry out of the darkness of the city street. In both stories, the protagonists are entirely and completely clueless as to what they have done to deserve such mysterious justice; indeed, that willful ignorance is itself their fundamental crime. And, of course, they are both condemned, as also are readers who have colluded in their willful denial of guilt. These novels would have depressed even Thomas of Celano, the author of the "Dies Irae" poem, so long part of funeral liturgies.

During the twentieth century there were attempts to avoid juridical language and reformulate this sense of taint or lack or twistedness right at the heart of human experience with death. Miller-McLemore's is one excellent example of this, as she clearly identified the problem. Try to express this sin-death connection, minus the juridical language; it is harder than we think. If we insist that biologically death is "a natural event," we evade the way in which for us it is always a negative reminder of finitude and lack, not a natural event in the way the year's cycle of seasons are natural. For example, we can focus on existential guilt rather than legally imaged guilt, on "before God we are always in the wrong" (this wording is Kierkegaard's, always an ominous sign), without needing to point up specific acts of wrongness. Karl Rahner, wrote of "death's darkness" as an alternative (Rahner 1961, p. 50). This phrase evokes a sense of fear and mystery that the traditional language of punishment and propitiation in the end obscures. Death's darkness does not mean we can never have a good dying, but that there is something un-good about death as humans experience it, and that this can never been completely erased or obliterated. It may mean that while death can be provisionally accepted, it can never be completely acceptable. Or, as Henri Nouwen famously stated, "In God there is no death" (Nouwen 1982, p. 75). And therefore, there is a limit to how one can befriend death, as Nouwen initially tried to do, or how use of natural imagery can overlay such a built-in sense of death's darkness.

Rahner retained a modified Platonism—death as "the separation of body and soul"—but unlike Seth Pringle-Pattison, he believed that death really does affect the soul, darken it, and seep into the corners of even the most triumphant Christian death, that of the martyr. The remedy for death's darkness is not just lots of more light, but awareness that for humans death is "a free act," as well as biological fate. We may grasp our deaths in advance as elements of our own stories, lived out by our active forming of our lives as totalities. Rahner—and anyone else who writes this way—is not dealing with particular empirical cases. The person who is killed suddenly when an airplane crashes down from the sky onto him, is not directly experiencing death as a free act, or indeed aware that his biography ends at that moment. On the other hand, Bunyan's pilgrims and Callanan and Kelley's hospice patients could all die deaths that Rahner would find illustrative of his basic theme. As I understand this, even the deaths of Christians, who consciously follow the model of Jesus into death and therefore "dieth well," will not overcome all of the darkness of death. In this age of the world, it remains.

And so, in the traditional rite of burial, the Book of Common Prayer includes this anthem:

> You only are immortal, the creator and maker of mankind, and we are mortal, formed of the earth, and to earth shall we return. For so did you ordain when you created me, saying "You are dust, and to dust you shall return." All of us go down to the dust; yet even at the grave we make our song: Alleluia, alleluia, alleluia. (The Episcopal Church 1979, p. 499)

6.5 What About Mourning?

In all the focus on Jesus' death and resurrection, there has also been a noticeable lack of attention to bereavement. What follows death, for Christians, is the hope of resurrection, something that happens after to the person who has died. There is in Christian traditional rites of funerals little or no attention to the other persons for whom the death matters. Their ruptured lives and sense of loss have been there but never theologically attended to, nor integrated into the overall perspectives on Christians and death. The people at the funeral were addressed as the future dead, so to speak, not primarily as current grievers. While Protestant preachers at funerals allowed that grief was universally human, to focus on it distracted attention to the real message of the Gospel. Grieving had no religious significance in and of itself (Daniels 1937; Blackwood 1942). That did not mean, of course, that patterns of mourning, and the social roles assigned to mourners, were not real and powerful. But they were cultural, not theological concerns. This in contrast, say, to Jewish *shiva*, which is tied to the Torah by very specific injunctions.

There is, however, a partial exception to the omission of grieving, unnoticed by Protestants but central at least in the past to Roman Catholics. This is the motif of the Pièta, of "Stabat Mater," Virgin Mary's grieving the death of her son. Michaelangelo's famous sculpture is of the gigantically proportioned bereaved mother who holds the dead body of her adult son, who appears passive and less powerful. No, I do not accept Julia Kristeva's eccentric declaration that this figure is *the* central one of Christianity (Kristeva 2002). But it is certainly a representation of grieving.

However, the primary place of bereavement has been in Lenten and Good Friday devotions. Here, especially within the Stations of the Cross as traditionally organized, we find mourning Mother yet also the centrality of Jesus' own dying and death. The poem "Stabat Mater" is worth quoting here.

> At the Cross her station keeping
> Stood the mournful Mother weeping,
> Close to Jesus to the last.
> Through her heart His sorrow sharing
> All His bitter anguish bearing,
> Now at length the sword has paased. (da Todi n.d.)

This poem's scene is John 19:25–27, when Jesus' mother and the beloved disciple remain to watch him die when all others have fled. The reference to the piercing sword harkens back to a prophecy early in Luke's Gospel, (Luke 2:35), when Mary is given forewarning of the death of her son. The piercing of Mary's heart may also have been linked to the piercing of Jesus' own corpse by the Roman soldier who removed it from the cross (John 19:34), at least by the author of the poem. Once again, I should emphasize that mourning as such is not the focus. Indeed, "Through her heart His sorrow sharing" links Mary not to post-death bereavement but to Jesus' own agonized dying.

Another thing to note is that the sorrow of everyone else in the story is overlooked, even that of the Beloved Disciple who promises to care for Mary. As a male figure, he is side-lined in this very gendered portrayal of grief. In our own day, this focus on maternal mourners has continued, perhaps as an echo of the Stabat Mater symbol; here is where Julia Kristeva may have gotten things right. Organizations of bereaved mothers protest wars and political tyranny, mobilizing passionate grief into anger against those who caused the deaths of their sons. Examples from Argentina, the Ukraine, and Israel show the power of this imagery, its unsurprassed moral claim. In our country, it is Mothers Against Drunk Driving (MADD) who has organized to tighten laws about highway safety. As has been pointed out at a recent conference sponsored by the Center for the Study of Death and Society (2017), there are no such organizations or mobilizations of grieving fathers.

Now, I have not found that the Stabat Mater has figured in twentieth-century or twenty-first century funeral preaching, in contrast to its traditional place in Good Friday piety. As already noted, this theme is so coded as Catholic, that no Protestants are likely to mention it, as even those of us non-Catholics who have practiced the Stations of the Cross have substituted less Marian poems and readings for Jacopone's poem. As for contemporary Roman Catholics, the explicit theological theme is expected, at funerals, to remain the death and resurrection of Jesus Christ. (See the Most Rev. John Myers 2003 for a particularly stringent statement of this policy.) Mary is as side-lined here as are all biographical anecdotes about the deceased. As for funerals for dead children, I cannot say that either Catholics or Protestants or anyone else today would find the "Stabat Mater" in this context; the terrible grief of burying a child does not, for us, call forth this particular image.

While there have been contemporary prayer and worship manuals to allow more space for grief, it is still very hard to find a place for loss and lament in Christian thought and practice. The emphasis on resurrection as the sequel to death squeezes out these as theological if not pastoral possibilities. Were there other sources for a genuinely Biblical vision of mourning, as a meaningful religious experience? For some scholars, the genre of lament poem is such an alternative. Laments in the Hebrew Bible, especially found in Psalms, often expressed outrage, not passive resignation or mournful weeping as a state in itself. Those contemporary organizations of bereaved mothers clamoring for justice resonate with this tradition, even more than with the Pièta. But did laments such as in the Psalms appear in the New Testament? Not really. And are all elements of lament really suitable for Christians? The scope of Biblical laments included the cry for God to smash your enemies' children to death against the rocks. (C.S. Lewis, in his writing on the Psalms [1958], has to take up this question, giving a rather evasive answer compared to that of Biblical theologian Walter Bruggemann [2000]). As with the debate over holism vs. dualism, the question of lament and the quest for justice has been intertwined with a set of contemporary concerns over the Holocaust, genocide, and political oppression. While there may indeed be situations that are genuinely lamentable, or lamentworthy, they remain only partly thematized in Christian practice, let alone doctrine.

6.6 Hell vs. Progress

There is one traditional, debated issue that has gone unmentioned so far, in part because unlike holism vs. dualism, it seems very far from applications to medicalized death. That issue is the doctrine of hell, or of eternal separation from God. Clearly taught in the Bible (no one truthfully denies this), is this a necessary part of Christian teaching? Does one say "I believe in hell" in the same way that the creeds say "We believe in one God"? To overtly and directly question something so obviously Biblical was not to be publically tolerated until the early modern period. Origen's scheme of universal salvation, including the eventual salvation of demons, was condemned and that settled things, officially, for a long time. Dante Alighieri's magnificent poem "The Divine Comedy" (Dante 1961) made hell vivid and psychologically credible, so that his was the standard portrait of it for all Christians and non-Christians, down to today, along with his vividly depicted Purgatory and Paradise. In Dante, the damned are locked into themselves, their own worst enemies as they were in life. "They yearn for what they fear" (*Inferno* III, 23). Demons are mostly adjuncts. The horrors of hell are not those of a Hollywood supernatural horror film, although Hollywood has tried and tried.

By the seventeenth century, however, questioning of church doctrines was more open, at least in some circles. Once any thinker began to defend the traditional teaching on eternal condemnation on any rational grounds, beyond just saying "It's in the Bible," there were basically two arguments in its favor:

1. Divine justice demands it. When we offend God through sin, we offend that which is infinite—unlike a crime against a mortal, limited being. Therefore, an infinite penalty is just. What "infinite" means in this context is debatable, but it certainly includes never-ending. The penalties of hell go on and on and on forever.
2. The second argument is that hell serves as a deterrent to sin. People who would otherwise commit serious sins will refrain, afraid of the ultimate, eternal consequences.

Both of these "rationalist" arguments started to crumble in the early modern and Enlightenment era. Daniel Walker's *The Decline of Hell*, takes us through this. So long as punishment's purpose was retributive justice, the first argument made a certain amount of sense. But if the purpose of punishment at the human level becomes rehabilitation, then endless punishment is pointless. No one in hell ever repents, learns their lesson, and changes. In Dante's portrayal of hell, the damned just feel anger and self-pity, not true repentence. So it was changing ideas about human justice that made hell appear cruel and futile, rather than a symbol of God's eternal justice. When this happened, the weakness of the second argument became apparent too. Most of us admit we aren't perfect and may deserve some punishment for our sins. But eternal torment? This is excessive and unbelievable as an actual motivator. A witty French woman named Marie Huber commented that "the fear of a violent distemper of twenty or thirty years continuance" would make a more effective, psychologically believable threat (Walker 1964, p. 41). This is, I believe, the same kind of debate that went on for years in our era about the validity of capital punishment as a deterrent. Note, however, that in Dante, hell does not serve as a deterrent in this sense. He experiences the pains of damned sinners in order to recognize and repudiate sin in himself. Sin on display in all its sinfulness, stripped of any trace of glamour or titillation or glory, is what hell as a deterrent is really about, not just fear of future consequences.

These arguments continued, not very productively, through the nineteenth century especially. How could a God of love rule over such a torture-realm? How could a God of justice not avenge wrongs, when the voices of innocent blood cry out for recompense? But one nineteenth-century argument seemed to make hell obsolete. That was belief in Progress. Progress seemed to be the law of nature, of humanity, of the universe. When nineteenth-century intellectuals thought of evolution, whether they accepted Darwin's version or not, they thought of Progress. When they looked around at marvels of engineering, such as railroads and telegraphs and steel bridges, they saw Progress. When they looked at urban slums, they saw the need for Progress. And when rapidly colonizing far-off places of the world and their peoples, they justified this in terms of Progress. Those natives, left to themselves, were never going to progress beyond savagery, but with Europeans' help, they too could experience Progress. And hell was the realm of no progress. "Abandon hope, all ye who enter here" was therefore an affront to a universal law. It was intolerable that such a place could exist at all.

I have capitalized Progress, because, retroactively, this seems an idolatrous belief. Better sanitation and higher rates of literacy, job safety and security, and the protection of children are all wonderful goals, but they do not make a cosmic law. Contemporary biology's understanding of evolution is not tied to Progress, although it is often hard to disentangle these for us non-scientists. But when Progress was divinized, Christian imagery of eternal existence had to change too. Heaven became a realm of perpetual growth and learning, not just rest and worship (Bregman 2011, p. 55). That seemed too static, although Dante's Paradisial residents do not seem bored by perpetual contemplation of the Divine. And if heaven is a place of growth and change, then hell as traditionally conceived has no place at all, unless it is—following Origen?—transformed into a really tough rehab program. It would thus resemble purgatory, a realm Protestants had rejected because it is not clearly Biblical.

Indeed, if Progress is the true cosmic law, then static eternal realms of either the New Testament or Dante should give way to a more dynamic vision of the afterlife as a destination. This, I believe, has been the basic appeal of reincarnation, as Westerners learned of it and appropriated it. Whatever Hindus and Buddhists taught, the nineteenth-, twentieth- and twenty-first-century Western version of reincarnation is that it establishes perpetual learning and growth as the destiny of all souls. We probably need more than one lifetime to do this thoroughly, but over a sequence of these, we will all experience Progress. Advocates of reincarnation have relied on this hope and attacked the traditional Christian eschatology for its absence of Progress. To quote famous nineteenth-century occultist philosopher Allan Kardec, "Birth, death, rebirth, and progress without end, this is the law" (Sharp 2006, p. v).

It was therefore one of the goals of early twentieth-century theology to try to deflate this idol of Progress. World War I was probably a more powerful deflator. Moreover, hell wasn't so easily abolished from Western imagination, whether Christian or not. Once again, literature by non-Christians shows it is alive. Malcolm Lowry's *Under the Volcano* (1947) is about several characters' travels through a landscape, which one of them at the end recognizes as hell, although the title tells us this right from the start. (Under Mt. Aetna was a traditional entrance to Tartarus, Roman hell.) Jean-Paul Sartre's play "No Exit" (1946), even better known, has as its message: "Hell is other people." Tragically and horribly, it appears that those who designed concentration camps deliberately modelled them on traditional hell (DesPres 1976, p. 94). "Abandon hope, all ye who enter here": they build an environment to make that compellingly real for the prisoners.

In the light of this, some thoughtful Christians and non-Christians have reassessed the meaning of hell. Not to bring it back as a literal place, let alone as a deterrent in the traditional manner. But maybe to acknowledge that some imagery of utter separation and hopelessness is needed for a complete picture of the human condition. Even when excluded for reasons of humanistic compassion, hell reappears on the map. I hesitate to go further than this, but such a possibility counters the knee-jerk reaction that Progress forbids us to go there. God may forbid, but Progress is an idol, not a universal cosmic law.

6.7 A Larger Setting for Dying and Death

In 1973s classic *The Denial of Death*, Ernest Becker found that the point of religion was not to deny death, but to open a vision of the human condition on the largest possible stage. We need "hero systems," and yet they are all, every one of them, partly illusion. We can never stand outside the human condition, but there is a way to vision it with a certain tragic grandeur that many ideologies, both religious and psychotherapeutic, timidly avoid (Becker 1973, p. 276ff.). The death and resurrection of Jesus, as cast in Becker's light, certainly attempts such a cosmic/heroic framing of death's meaning for humans. While I do not support Becker's attempt to turn Freudianism into such a tragic, transcendent system, we may allow that this larger stage has been a strength of Christians' encounter with death, although often overlaid with promises of hope that Becker would find ludicrous. To harken back to Bunyan's allegory of pilgrimage, that river at the end of life is there for all of us. It is not for the faint-hearted, and the systems of easy denial and accommodation that Becker targeted resemble the ferry used by Ignorance to get across it. Without endorsing the major arguments of Becker, whose Freudian misogyny was offensive even back in 1973, he suggests how puny and small-scaled many of the contemporary perspectives on psyche and on death really are.

This is the basic contention, I believe, that Christians should mount against medicalized death. Its stage and the drama over certain issues such as physician-assisted dying or abortion is not wrong in the sense of morally flawed. It is simply too confining. Those persons who wrote their autobiographies in the wake of the Death Awareness Movement recognized this. A story built around blood-test results is demeaning to both author and readers. A life whose choices are dictated by doctors' expertise and treatment options misses what really matters most to human beings. Medicalized death may have become for us another variant of that belief in Progress which found hell unacceptable. If diseases can be conquered, then keep trying until we can cure death itself. Against the reality that aging inevitably breaks down the body's resistance to disease and malfunctions, there seems to float the hope that we can live death-free, a hope denounced by Sherwin Nuland's *How We Die* (1993) and Atul Gawande's *Being Mortal* (2014). We seem to have transformed basic existential questions into questions over technology. "If we can, we must" says at least one critic of this stance. But it is not even the stance itself; it is the entire frame of reference that is diminished and diminishing. We need larger dramas, more expansive and imaginative categories, than what medicine can provide.

The paradox is that Christianity has traditionally tried to offer this through the particularities of one person's historical dying and death. While Becker and many others, both in and out of Christianity, prefer a universalizing framework, whether Plato's or anyone else's, Christians have kept relentlessly focused on the death and resurrection of Jesus. This story is what Christianity starts from; in that respect Cullmann was absolutely right. A seemingly more universal-philosophical starting point, such as sometimes attributed to Buddhism, would not be true to this tradition and its messages. This particularity does not mean that "one size fits all" squeezed all actual Christians' dying into a mold; nor has it eliminated the need to rely on some philosophical categories to explore the implications of Jesus' death for humanity. I do not find that the tradition's limits as a resource for persons today are due to its intense focus on one Savior's death, however unique and empirically unusual that death. Its limits are certainly present, and we have explored some of these. Yet it is our own limits of thought and imagination that Christian understandings of death can illuminate.

As for that glossy Christian book club brochure with no books on death, it may have reinforced the dominance of medical categories, by seceding territory to medicine while eagerly challenging popular secular psychologies of parenting and family life. In fact, that brochure stands as evidence for the lack of courage in learning from a tradition whose central theme includes death. Maybe not all of the past can be immediately appropriated as a model for us, but lack of interest in even trying is, I now believe, pathetic and incredibly unhelpful. A Christian response to the limits of medical understandings could have been offered, ought to have been offered, not at the level of learned scholarly theology, but in the same genre as the majority of books advertised by that club—a response that takes medicine and its limits seriously, but that moves all of us into asking other questions.

Whether one crosses the death-river of Bunyan's allegory in fear or confidently, this is not and never will be reduced to a medical question. Bunyan at least was willing to allow that both attitudes could be compatible with genuine faith. What is incompatible is to seek that ferry ride, chosen by Ignorance, that seeks to avoid individual crossing-pain altogether. In ways Bunyan himself never imagined, that is what medicalized death has offered. And in the death of Jesus and his resurrection, Christians could discover and rediscover ways to move beyond this. In the words of the Passion Choral hymn, "Who dieth thus dies well."

References

Ariès, Philippe. 1974. *Western attitudes toward death*. Baltimore: The Johns Hopkins University Press.

Aulén, Gustaf. 1967 [1930]. *Christus victor*. Trans. A. G. Herbert. New York: Macmillan.

Becker, Ernest. 1973. *The denial of death*. New York: The Free Press.

Blackwood, Andrew. 1942. *The funeral: A sourcebook for ministers*. Philadelphia: Westminster Press.

Bregman, Lucy. 1999. *Beyond silence and denial: Death and dying reconsidered*. Louisville: Westminster John Knox Press.

———. 2011. *Preaching death: The transformation of Protestant funeral sermons*. Waco: Baylor University Press.

Bruggemann, Walter. 2000. *Deep memory, exuberant hope: Contested truth in the post-Christian world*, ed. Patrick Miller. Minneapolis: Fortress.

Bunyan, John. 1964 [1678]. *The pilgrim's progress*. New York: New American Library.

Callanan, Maggie, and Patricia Kelley. 1992. *Final gifts: Understanding the special awareness, needs and communications of the dying*. New York: Poseidon Press.

Camus, Albert. 1956. *The fall*. Trans. Justin O'Brien. New York: Vintage Books.

Center for the Study of Death and Society. 2017. *Conference: Death at the margins of the state*. UK: University of Bath. June 9 10, 2017.
Cooper, John. 1989. *Body, soul and the life everlasting*. Grand Rapids: Wm. Eerdmans.
Cullmann, Oscar. 1965. Immortality of the soul; or, resurrection of the dead. In *Immortality and resurrection*, ed. Krister Stendahl, 9–53. New York: Macmillan Co.
da Todi, Jacopone. n.d. *Stabat mater*. Trans. Fr. Edward Caswall. www.preces-latinae.org/thesaurus/BVM/SMDolorosa.html.
Daniels, Earl. 1937. *The funeral message: Its preparation and significance*. Nashville: Cokesbury Press.
Dante Aligheri. 1961. *The divine comedy*. Trans. John Ciardi. New York: New American Library.
DesPres, Terrence. 1976. *The survivor: An anatomy of life in the death camps*. New York: Pocket Books.
Gawande, Atul. 2014. *Being mortal*. New York: Henry Holt and Co.
Gerhardt, Paul. 1656. O sacred head, now wounded. 1820 (English translation: J. W. Alexander of "O Haupt voll Blut und Wunden.")
Gibson, Mel, and Benedict Fitzgerald. 2004. The passion of the Christ. 2005. Directed by Mel Gibson. Icon Productions.
Gusmer, Charles. 1984. *And you visited me: Sacramental ministry to the sick and the dying*. New York: Pueblo Press.
Kafka, Franz. 2015 [1925]. *The trial*. New York: Tribeca Books.
Ketcham, William, ed. 1899. *Funeral sermons and outline addresses: An aid for pastors*. New York: Harper and Bros.
Krieg, Robert. 1984. The funeral homily: A theological view. *Worship* 58: 222–239.
Kristeva, Julia. 2002. Stabat mater. In *The portable Kristeva*, ed. Kelly Oliver, 310–332. New York: Columbia University Press.
Lawrence, D.H. 1959 [1929]. *The man who died*. New York: Penguin Random House.
Lewis, C.S. 1958. *Reflection on the psalms*. New York: Harcourt Brace.
Lewis, Alan E. 2001. *Between cross and resurrection: A theology of Holy Saturday*. Grand Rapids: Wm. Eermans.
Lowry, Malcolm. 2007 [1947]. *Under the volcano*. New York: Harper Perrenial Modern Classics.
Miller-McLemore, Bonnie. 1988. *Death, sin and the moral life*. Atlanta: Scholars Press.
Myers, David. 1978. *The human puzzle: Psychological research and Christian belief*. New York: Harper and Row.
Myers, John J., the Most Reverend. 2003. Reports on policies for funeral liturgies need clarification. Roman Catholic Archdiocese of Newark. http://www.rcan.org/archbish/jim_articles/sit103-02-05.htm. Accessed 05 Feb 2003.
Nouwen, Henri. 1982. *A letter of consolation*. San Francisco: Harper & Row.
Nuland, Sherwin. 1993. *How we die*. New York: Vintage Books.
Paxton, Frederick. 1990. *Christianizing death; the creation of a ritual process in early medieval Europe*. Ithaca: Cornell University Press.
Pelikan, Jaroslav. 1961. *The shape of death: Life, death, and immortality in the early fathers*. New York: Abingdon.
Rahner, Karl. 1961. *On the theology of death*. Edinburgh: Nelson.
Sacks, Oliver. 2015. My own life: Oliver Sacks on learning he has terminal cancer. *New York Times*. www.Nytimes.com/2015/02/19/opinion/oliver-sacks-on-learning-he-has-terminal-cancer.html. February 19.
Sartre, Jean-Paul. 1989 [1946]. *No exit and three other plays*. Trans. Stuart Gilbert. New York: Vintage.
Schnackenberg, Rudolph. 1964. *Baptism in the thought of St. Paul*. Trans. G.R. Beasley-Murray. London: Herder and Herder.
Seth Pringle-Pattison, A. 1922. *The idea of immortality*. Oxford: Clarendon Press.
Sharp, Lynn. 2006. *Secular spirituality, reincarnation and spiritism in nineteenth-century France*. Lanham: Rowman and Littlefield.

The Episcopal Church. 1979. *Book of common prayer*. New York: The Church Hymnal Corporation.
Thomas of Celano. n.d. Dies Irae. In the Requiem Mass of the 1962 *Roman Missal*. English translation at: https://en.wikipedia.org/wiki/Dies_irae.
Urban, Linwood. 1995. *A short history of Christian thought*. New York: Oxford University Press.
Verhey, Allen. 2011. *The Christian art of dying: Learning from Jesus*. Grand Rapids: Wm. Eerdmans.
Walker, Daniel. 1964. *The decline of hell*. Chicago: University of Chicago Press.
Wesley, John. 1981. A plain account of Christian perfection. In *John and Charles Wesley: Selected prayers, hymns, journal notes, sermons, letters and treatises (Classics of Western spirituality)*, ed. Frank Whaling, 342–347. New York: Paulist Press.
Wolfson, Harry. 1965. Immortality and resurrection in the philosophy of the church fathers. In *Immortality and resurrection*, ed. Krister Stendahl, 54–96. New York: Macmillan Co.

Chapter 7
A Jain Ethic for the End of Life

Christopher Key Chapple

Abstract Jainism, which arose in India more than 2500 years ago, states that the soul is eternal: it has never been created nor can it ever be destroyed. The soul becomes cloaked, birth after birth, with karmas that obscure its true nature. The utmost task for the human being entails purifying oneself of karma through untying its many knots that bind the soul, masking its innate energy, consciousness, and bliss. One technique to guarantee a better life in the next birth is to die a conscious death through a systematic process of fasting, entering into a state of dehydration. This highly regulated practice, pursued by monks, nuns, and laypersons who have gone through a rigorous period of internal reflection and external assessment before embarking on this path, provides a peaceful way to embrace death. Known as *salle-khana* or *santara*, it has recently been challenged in the courts as a form of suicide, an illegal practice, though for the Jain community it remains an important option through which one can express religious faith.

7.1 Jainism: An Ancient Religion of India

The Jain faith arose before Buddhism in the northeast area of India. Its most recently lauded, liberated teacher, Mahavira Vardhamana, lived at the same time as the Buddha, approximately five centuries before the birth of Christ according to tradi-tion. He is said to be the 24th in a line of great teachers or Tirthankaras. These individuals have crossed from the realm of bondage into a place of abiding freedom, from *saṃsāra* into *kaivalyam*, the Jain equivalent to Buddhist *nirvāṇa* or Hindu *mokṣa*. A liberated Jain resides eternally in a state of energy, consciousness, and bliss, perched high above the universe in the Realm of Spiritual Success, the Siddha Loka. In addition to the 24 great teachers, an unknown number of other liberated souls dwell therein, known as the perfected ones or Siddhas. Followers of the Jain faith seek to emulate the Tirthankaras and the Siddhas, who have ascended into the

C. K. Chapple (✉)
Loyola Marymount University, Los Angeles, CA, USA
e-mail: cchapple@lmu.edu

T. D. Knepper et al. (eds.), *Death and Dying*, Comparative Philosophy of Religion 2, https://doi.org/10.1007/978-3-030-19300-3_7

realm of freedom, as well as three groups of living exemplars: the leaders of religious orders, teachers of the faith, and the monks and nuns who have taken the requisite vows to advance on the path of freedom, the *mokṣa mārga*. One vow that supports this pilgrimage is the fast unto death.

7.2 The Soul

Jainism posits that the soul cannot be killed. The soul has been described succinctly in the Bhagavad Gita, a Hindu text also revered by Jains:

> II:20: [The soul] is never born, nor does it die;
> nor, having come to be, will it not be once again.
> Unborn, eternal, everlasting,
> this primal one is not slain when the body is slain.
> II:25: It cannot be cleaved in two, burned, wetted, or dried.
> It is eternal, all-pervasive, unchanging, and immovable.
> II:26. Even if you think that this soul is perpetually born and perpetually dying,
> even so, Strong Armed One, you should not grieve for this.
> II:27: For to one born, death is certain.
> And to one dying, birth is certain.
> Therefore you must not grieve over what is unavoidable. (deNicolas 2004, pp. 33–34)

Unlike most Hindus, the Jains define the soul as having two aspects: an eternal sense of self-awareness, and a cloak of matter that obscures self-awareness. The terms used by the Jains are *jīva* for the soul and *karma* for the material cloak. The word *jīva* comes from the verb root *jīv* which means to live. Unlike the Hindu word for soul, *ātman*, which derives from the root *āt* which means to breathe and implies that this breathing takes place in a human body, the Jain reference to life and living is more expansive and more inclusive. According to Jainism, anything that vibrates (*spanda*) carries life force, from miniscule elemental particles to large trees and mammals. This life force always manifests into an individual presence. Through its actions, it attracts matter (*pudgala*) in the form of specific karmas that are sticky and colorful. As these karmas cloak the soul, they obscure its innate energy, consciousness, and bliss, binding the soul like creeping tendrils hide a tree. Through adherence to vows, one can slow the karmic creep and eventually allow the karmas to disperse. One achieves freedom through the dispersal of karmas.

The Jain worldview asserts the reality of both tree and tendrils, soul and karma. Jainism also teaches that there is no singular universal soul, only an infinite number of souls, present from beginningless time, seeking to survive and thrive, and in the case of a few, seeking to obtain freedom from karma. This philosophy differs radically from monist schools of Hinduism which seek union with the universal soul and from the teachings of Buddhism, which deny the existence of a soul.

The two primary branches of Jainism, the Digambaras and the Svetambaras, agree on these basic teachings. Due to famine around 2300 years ago, some Jains had left the region of northeast India and settled in the south and central part of India. This group, the Digambaras, expects its senior monks to embrace the vow of nonpossession or *aparigraha* to the fullest, renouncing all clothing. The group that headed to the western part of India, the Svetambaras, expects monks and nuns alike to remain covered with white garments. Their respective names mean "clad in space" and "clad in white."

7.3 Vows

Five core vows define the life of a Jain, whether a layperson or a monastic. These vows require adherence to the precepts of do-no-harm (*ahiṃsā*), tell-no-lies (*satya*), do not steal (*asteya*), remain sexually chaste and if married remain faithful to your spouse (*brahmacarya*), and minimize possessions (*aparigraha*). For monks, nuns, and people within families these principles govern not only everyday behavior but one's outlook toward life and death. For a Jain, the purpose of life is clear: to advance through one's ethical understanding and practice toward greater states of purity, halting and dispelling the karmic cloak. Furthermore, because of the Jain view on the eternal nature of life in its myriad forms, the Jain outlook toward death seeks not to avoid or deny death, but to build a life that embraces and even invites the inevitability of death. Death is not seen as an unknown abyss but as a transition into another expression of life, whether taking a new body in the realm of *saṃsāra* or ascending to the perfect freedom of *kevala* in the Siddha Loka.

The five vows described above are common to Hinduism through Patanjali's Yoga system and to Buddhism through the five precepts, though the Buddhist replace non-possession with a mandate to abstain from all intoxicants. Perhaps the most distinctly Jain vows are nonviolence and non-possession, *ahiṃsā* and *aparigraha*. Jains are well known for their adherence to nonviolence. They inspired many to adopt vegetarian diets and to stop making animal sacrifices, and they inspired Mahatma Gandhi as he developed his program of campaign for India's liberation from British rule, Hind Swaraj.

For the purposes of understanding the Jain approach to death and dying, we will focus in this chapter on the Jain commitment to *aparigraha*. First, a Jain practices *aparigraha* in service of nonviolence. The less one owns, the less harm one commits. Second, monks and nuns are held to a very high standard in regard to minimizing possessions. They are only allowed to own what can be carried. Monks and nuns renounce access to kitchens, surviving only on donated meals.

In some cases, this means eating only food directly place in their hands. Twice each year, Jain monks and nuns pluck out their hair, shedding any mark of vanity. Svetambara monks and nuns are allowed only one change of apparel. Digambara monks renounce even clothing as they move into advanced states of asceticism. To own less means that one becomes more unencumbered and hence closer to freedom.

Aparigraha extends to giving up attachment to body itself. In Yoga philosophy, the karmic impulse to persist, *abhiniveśa*, impels one to hold onto life as long as possible and, following death, to find a new vehicle for experience. The Upanishads provide a biological narrative to explain the journey of the bundle of unrequited desires released in the funerary process as it moves from the cremation fire to the atmosphere, mingling with the clouds, eventually coming to earth in rainfall, nourishing plants that become food that then is processed into ovum and sperm, allowing a new pathway for life. In Tibetan Buddhism, the subtle body comprised of unresolved karmas travels for 49 days before entering a new body.

The afterlife narrative in Jainism folds back into the specific deeds performed in the final third of life which program the exit of body. Rather than traversing a sequence of agricultural cycles or enduring a 49-day intermediary period, the Jain tradition claims that a very specific and efficient group of karmas propel the soul immediately into a new embodiment. The karmas, known as the *āghātiya* or the non-destructive group, consist of name and form, family, feeling, and life-span karmas. According to Padmanabh Jaini:

> A person's life-span karma is fixed or bound only once in a given lifetime…. [T]his event takes place sometime during the last third of that lifetime and the individual in question is never aware of its occurrence…. By earnestly adhering to the path of proper conduct, a Jaina can hope, during the latter portion of life, to great influence the determination of life-span karma and thus the character of the entire next existence. (Jaini 1979, p. 126)

The Jain bodily mechanism consists of a karma body connected to an energetic body. This energetic body *(taijasa-śarīra)* "houses the soul as it leaves the body" (Jaini 1979, p. 126). Jaini explains the transitional moment as follows:

> A soul is said to be inherently possessed of a great motive force. Set free from the state of gross embodiment, it flies at incredible speed and in a straight line to the destination with its accompanying karma has deemed appropriate. This movement … is said to require … only a single moment in time, regardless of the distance to be traversed…. [T]he soul moves to its new state of embodiment in a straightforward and virtually instantaneous manner. (Jaini 1979, pp. 126–127)

Because of attachment, one takes on a new form instantly in the next moment after death. Monks and nuns, through their careful cultivation of nonattachment through the relinquishment of possessions, hope to ideally avoid rebirth altogether and ascend directly to the Siddha Loka. Most likely, however, their asceticism will enable them to attain an auspicious birth through which they may resume their quest for purification.

7.4 The Final Fast

One's attitude toward death according to Jainism can be best informed by relinquishment of all attachment, even attachment to nutrition. Voluntary death through fasting, known as *sallekhana* or *santara*, is pursued in the Jaina tradition when death is imminent due to disease or when one is unable to function self-sufficiently. To hasten death by ingestion of poison or some other form of self-harm would be an act of violence and not acceptable. Jains prepare for death by fasting intermittently throughout life. This allows one to develop the fortitude to face the final fast, without fear the ultimate act of renunciation. In the case of Jaina monks or nuns or even laypersons, the final fast might commence when one is no longer able to abide by monastic rules governing nonviolent behavior due to the debility of old age or infirmity. The fast unto death generally takes place at the close of a normal lifespan. It would be unacceptable for a young, healthy person to enter the final fast.

This chapter will summarize the central features of the Jain tradition of dying. It will then discuss Dr. Sherwin B. Nuland's observations on the death process in contemporary America in contrast with three North American instances of fasting to death: one undertaken by an observant Jaina laywoman, the other by Scott Nearing, a noted American radical, and the third by the husband of a radio journalist. The chapter will close with reflections on longevity, expectations for a life well lived, and how the Jain practices in regard to death, despite being legally challenged in India, can contribute to the global conversation on dying and death.

For the Jain community, fasting to death celebrates a life well lived and emphasizes key aspects of Jain philosophy. First, it demonstrates a willingness to devote oneself in an ultimate sense to the observance of nonviolence. By not eating, no harm is done to any living being. Second, fasting burns off residues of karma that otherwise would impede the soul and cause further bondage. It purifies the soul by releasing the fetters of past attachment.

The tradition of fasting to death in the Jaina tradition has been documented extensively, most notably in the works of Collete Caillat (1977); Tukol (1976); Padmanah Jaini, whose book *The Jaina Path of Purification* (1979) opens with a description of the fast unto death of Muni Śāntisāgara, the spiritual preceptor of his childhood community in India; Shadakshari Settar, whose books *Inviting Death* (1986) and *Pursuing Death* (1990) document scores of cases from epigraphic records; my own writings on the final fast of a Jain nun in *Nonviolence to Animals, Earth, and Self in Asian Traditions* (Chapple 1993, pp. 104–105); Lucy Bregman's edited collection on *Religion, Death, and Dying* (2010); and Whitny Braun's 2015 doctoral dissertation, "Sallekhana: The Philosophy, Ethics, and Culture of the Jain End of Life Ritual;" to name a few resources.

The Digambara community refers to this practice as *sallekhana,* which literally means the thinning out of existence, from the verb root *likh,* which means to scratch or scrape. The Svetambara community tends to refer to this practice as *santara* (Babb 1996, p. 60), which translates as passing over or crossing, and comes from the verb root *tṛ,* related to the English word turn. Each community has developed careful rules for gaining permission to start a final fast. It cannot be undertaken unless approved by the community and, in the case of some Digambara renouncers, one must leave one's own monastic group *(gaccha)* and conduct the fast with the assistance of some monastic community that agrees to facilitate the process (Settar, personal communication, Australia 1994).

The most frequently quoted text cited in regard to the Jaina process of fasting to death was written by Samantabhadra in the second century of the Common Era. Known as the *Ratnakaraṇḍaka Śrāvakācāra,* it states:

> One should give up gradually all solid foods, increase the taking of liquids like milk, then give up even liquids gradually and take warm water. Thereafter, one should give up warm water also, observe the fast to the best of one's ability with determination and depart from the body repeating the *namskara mantra* continuously until the last. (in Tukol 1976, p. 8)

This formula of progressive withdrawal of nutrition begins with stopping solid foods, then caloric liquids, transitioning eventually from water to nothing. The death that follows comes from dehydration, not starvation. The organs cease to function, leading to the final exhale.

7.5 Death in Contemporary America

In *How We Die: Reflections on Life's Final Chapter,* Dr. Sherwin B. Nuland of Yale University Medical School tells the story of Miss Hazel Welch, a 92-year old resident of a convalescent home in Connecticut. One day, Miss Welch collapsed; she was diagnosed with operable peritonitis. Initially, she refused the operation, stating she had been on the planet long enough and did not wish to go on. Dr. Nuland talked her into the operation, and though her chances of surviving the operation were one-in-three, she did in fact live through the surgery. Her recovery proved agonizing. She required a breathing tube for 9 days and, as Dr. Nuland writes, "she spent every minute of my twice-a-day visits staring reproachfully at me" (1994, p. 252). As soon as she returned to her convalescent home she arranged with her trust officer from the bank that handled her estate to draw up papers assuring that she would in the event of a future health failure receive no more than nursing care; "she wanted no repetition of her recent experience and emphatically said so in her written statement" (1994, p. 252). Two weeks later she had a massive stroke and died in less than a day. Somewhat ruefully, Dr. Nuland writes that he wished he had abided by her wishes the first time and agreed not to perform the initial surgery. However, he also notes that his colleagues of his hospital's weekly surgical conference would have disapproved, retorting with a remark such as "Does the mere fact that an old lady wants to die mean you should be a party to it?" (1994, p. 253).

Nuland points out that 80% of American deaths occur in the hospital. As French historian Philippe Aries has noted, "Our senses can no longer tolerate the sights and smells that in the early nineteenth century were part of daily life, along with suffering an illness…. [T]he hospital has offered families a place where they can hide the unseemly…. [T]he hospital has become a place of solitary death" (Nuland 1994, p. 255). Though he offers no easy solutions, Dr. Nuland presents the stark reality and pervasiveness of an alienated and fundamentally unhappy death process in America.

7.6 The Death of Mrs. Vijay Bhade

By contrast, I want to describe the death of Mrs. Vijay Bhade, a Jain woman suffering from sarcoma in West Virginia (see also Chapple 2010, pp. 205–206). She was raised within the Jain community in India and was married to Jain physician. Her struggle with illness led her to pursue treatment according to Western medical practices. She also applied an attitude toward death and dying learned from and encouraged by the philosophy of Jainism and the traditional practice of *sallekhana*. Modern cures were sought but when these proved futile, Mrs. Bhade actively pursued death in the traditional manner of gradually letting go, first of solid food, then liquids, then water. Her goal was to make a conscious transition into death. She died at home, surrounded with family and friends.

In an interview with Dr. Bhade, he commented that the passing of his wife was a beautiful experience. At age 43, stricken with sarcoma, she underwent 6 months of treatment to no avail. When it was seen that nothing more could be done, she explained to her three children (ages 17, 15, 13) that she was leaving. The last week of her life she took water and juice only. In the beginning of the week, she took a morphine drip for a time, but then decided to do without it; when she stopped the morphine, she no longer experienced pain. On the morning of her death, she called her friends and relatives at 4 a.m., asking them to come to the house. She took a bath and did puja. She asked for forgiveness of everyone (*kṣama*) and talked with her family members. She chanted several mantras including both the Namokara Mantra, which honors the great departed teachers and liberated souls, the living Jain leaders and teachers, and the nuns and monks, and the Samadhi Marana, a chant that welcomes death. Later in the morning she died alert and conscious. Those gathered were thrilled to witness the peacefulness of her passing (conversation with Dr. Vijay Bhade, October 1997).

In reflecting on the process of seeing his wife die, Dr. Bhade, as a physician, noted the differences between death in his Digambara Jaina community in Maharashtra and death in West Virginia. He observed that his exposure to death in India was somewhat limited, though, in his home community many choose to fast at what is deemed to be the end of their lives and that the *munis* or Jain monks take up the final fast when they can no longer keep up their vows. For instance, when their sight dims, the monks cannot effectively ascertain that no bugs have entered their food. Due to the difficulty in maintaining such basic practices of nonviolence, the monks will embark on a terminal fast. In the case of his own family, Dr. Bhade cited the instance of his mother. She experienced heart failure at the age of 74. She was offered but rejected angioplasty, and went on a liquid diet. Eventually, she entered a period of total fasting and gave up her life in a fully conscious state.

Many of the younger and middle-aged people have left West Virginia, seeking opportunities out of state. Consequently, many elderly are left behind without family in close proximity. The adult children responsible for the care of their elderly parents generally see them only once a year. Consequently, they are interested in prolonging the life of their parents (perhaps out of guilt, Dr. Bhade surmises) and will agree to extraordinary measures. As Dr. Bhade noted, "Terminal death is very painful. It can involve three to six months of torture. In some cases, families want to do everything possible [to keep the person alive].... Elderly people lying there so helpless with feeding tubes are a horrible sight. In general, in India, people do not suffer this way."

For religious, practical, and economic reasons, Dr. Bhade would support a greater awareness of the advantages of fasting to death. He stated at the onset of our conversation, "'There is an end to life.' This is the first step. People need to understand this.... When you are born, you are going to die." He suggests that the role of the physician is to make people comfortable, and that in many instances the prolongation of life with medical technology does not increase a person's comfort. In advocacy of fasting, he stated, "Fasting helps give up the attachment to this life. Desires grow less through fasting."

Dr. Bhade' s statements evoke basic Jaina cosmology. Desire, including the desire to live, can be an obstacle to one's ultimate happiness. By attenuating desire, one prepares to let go. By entering death in a process of conscious prayer, the transition, according to eyewitness accounts, becomes painless.

7.7 The Deaths of Scott Nearing and John Rehm

Scott Nearing (1883–1983), best known for his advocacy of simple living in the classic he co-authored with his wife Helen, *Living the Good Life,* chose an unconventional life. A pacifist and communist, he retreated to homestead in rural New England after he was dismissed from his professorship at the University of Pennsylvania due to his staunch opposition to World War I. With Helen Nearing he developed a maple sugar farm in Vermont, and eventually settled on the coast of Maine. The two pursued a life of learning and subsistence and managed to survive for several decades largely independent from the needs of the external economy. They grew their own food, built their own houses, and lived healthily on a vegan diet.

In his 99th year, Scott Nearing lost his physical mobility. After several months of near-total debility (LaConte 1997, pp. 16–21), he decided to stop eating. Helen Nearing describes the process as follows:

> My husband, Scott Nearing, died with deliberation and in full consciousness. He knew exactly what he was doing and planned it in advance. It was a death in keeping with his life—a reasoned process which he wanted to experience and make manifest. Death, to him, was merely the last stage in his growth, a natural organic act. He knew he was near the end and he wanted a death by choice—his own decision. His life had been sane and lived quietly and purposively. He wanted to go the same way—to live right into death.
>
> One day, as we were starting our evening meal at the table, he said, "I think I won't eat any more." He was 99, approaching his hundredth year. His body was wearing out, his usual vigor ebbing, and he was ready to call it quits. He thought he had lived long enough, and he was interested to know what lay beyond. He believed in some kind of survival of the spirit, some continuity, and was ready to enter into and partake of that phase as he had learned from and contributed to the physical phases of life
>
> From that time—a month before his hundredth birthday—he abstained from all solid food, taking only liquids. He waned and lost his strength, but kept his wits and good cheer—and determination. After a month of fasting on vegetable and fruit juices he announced, "Only water, please."
>
> His aim had been to avoid all pills, drugs, doctors, and particularly forced feeding in hospitals or nursing homes. He had wanted to die at home, quietly, and in his own good time. All of which he did, simply and serenely and resolutely. He stepped quietly and consciously into death.
>
> At the very end, alone with me, after a week on water, he breathed his last, detached and yet still aware. He drifted away and off, like an autumn leaf from the parent tree, effortlessly

> and tranquilly. It was a benign and calm departure—well-timed and appropriate. He breathed low; then he breathed no more. He went somewhere else, with active volition. He had practiced the art of dying well (Nearing 1995, pp. xi–xii).

The passing of Scott Nearing as told by his wife followed a parallel course to that pursued in Jainism.

Author Ellen LaConte is careful to point out that many hospice workers supported the final days of Scott Nearing. LaConte notes that

> Scott did not experience explicit organ collapse.... All Scott's systems ran down simultaneously until they ceased.... By the time Scott asked to go on juices...he had already been dying nearly nine months. When the decision to desist was finally enacted, he was attached to life by on the thinnest of threads.... He had no more interest in food than in world affairs.... Had he not stopped eating deliberately, he would soon have been unable. (LaConte 1997, p. 20)

LaConte also notes that both Helen and Scott Nearing had fasted routinely through their more than six decades together, preparing Scott well for his ultimate passing.

Such was not the case of John Rehm (1931–2014), a prominent Washington, D.C. lawyer who was diagnosed with Parkinson's disease in his late 70s. His wife, Diane Rehm, a prominent radio journalist, narrated his difficult passing in her 2016 book *On My Own*. She laments the lingering death of her husband. He suffered from Parkinson's disease for many years. "He no longer had the use of his hands, arms, or legs.... [H]e could no longer stand, walk, eat, bath or in any way care for himself on his own; he was now ready to die" (Rehm 2016, p. 3). He asked if there might be a way for him to gracefully end the pain and difficulty and was informed by his physician that his only option for escaping the pain of his illness was to refuse nutrition and hydration. The criminal code in Washington, D.C., forbids physician-assisted death. John Rehm decided to embark on a final fast after years of debilitation and 2 years in an assisted living facility. He stopped medication, food, and water on June 14, 2014 and passed on June 23rd. Both Diane and her husband John would have preferred to end his life medically but knew that death by dehydration was the only legal option. Similar to the cases cited above, Rehm was unable to sustain a reasonable quality of life and chose to invite death rather than wait for the inevitable, not unlike the path advocated by the Jains.

7.8 Legal Debates and Personal Choices

The Jain philosophy of life and death sees the two as part of the same continuum. Death results in life; life culminates in death, not as a finality, but as a transition. Jains, in preparation for this transition, fast on a regular basis, generally twice a month, and for a longer period during the Paryushan religious observances of late summer during which people seek forgiveness from others for misdeeds. At the end of one's life, the final act of expiation, the final sacrifice of one's body and karma, involves passing from this life consciously, at the conclusion of a successful period of fasting. By contrast, the drive to extend human life in contemporary medical practice, rather than allowing for a letting go, enforces a holding on to life that, for some, can be quite painful and distracting. In such circumstances, the inevitable passing into death becomes an arduous ordeal.

In India, celebratory events herald a Jain's decision to embark on a final fast. Advertisements like nuptial announcements or obituaries are placed in the newspaper and a general mood of happy anticipation can be detected. As noted earlier, the final fast is undertaken not only by monks and nuns but is also common among lay persons. In 2006, Nikhil Soni and Madhavi Mishra filed a Public Interest Litigation with the Rajasthan High Court arguing that *sallekhana* was suicide and hence an illegal act. They state that Article 21 of the Indian Constitution, which protects religious freedom, does not guarantee a right to death. They declared that *sallekhana* was used to coerce widows and elderly relatives to take their own lives. On August 10th, 2015 Rajasthan banned *sallekhana* and deemed criminal anyone who would abet the practice. On August 17, 2015, the Supreme Court of India put a stay on the ban temporarily (Braun 2015, pp. 205, 235). It is predicted that they will hold a formal hearing on this case by 2020.

Whereas in the United States there is a liberalization of attitudes toward death and dying, in India the case filed by Soni and Mishra seems to signal a desire to resist any willful move toward death. Deeming the fast unto death to be a form of suicide, the plaintiffs want to avoid any coercive action within Jain communities that might encourage undertaking the final fast. If the Supreme Court supports the definition that *sallekhana* is a form of suicide, then basic religious principles of Jainism would be brought into question. It is interesting that the work of Whitny Braun has been cited as providing historical evidence that to illegalize the fast unto death would violate the religious rights of Jains.

The young men who filed the brief hoped to end this practice in order to bring Indian culture into alignment with Western practice, arguing that human dignity becomes compromised when the end of life is seen as a transition to be embraced rather than avoided. Holding the practice of voluntary death in high regard, they argue, creates a social pressure in conflict with the greater common good. Two contemporary American writers, Diane Rehm as described above, and Barbara Ehrenreich, disagree with the notion that human life should be prolonged when no longer viable.

Barbara Ehrenreich, a cell biologist and print journalist, published the co-authored book *For Her Own Good: Two Centuries of the Experts' Advice to Women* in 1978, a history of how male authority defined and in many cases constrained women's health. Her latest book, 40 years later, returns to the topic of health, not from the perspective of a woman in the midst of bearing and rearing children, but from the perspective of an aging woman who has undergone treatment for breast cancer and, in an attempt to prolong her life, has committed herself to a thorough fitness regimen. The title of the book summarizes her current struggle: *Natural Causes: An Epidemic of Wellness, the Certainty of Dying, and Killing Ourselves to Live Longer*. From the onset, she makes clear that religion and spirituality hold no interest for her. She dismisses the notions of a soul and afterlife as unprovable and scoffs at practices such as Yoga that veer toward the religious. Nonetheless, she provides very insightful commentary on the contemporary state of the human condition regarding life, thriving, and dying. She had come to learn through her cancer experiences that the immune system, rather than curbing cancer, in some and perhaps many cases, actually enables the growth of tumors through large cells known as macrophages. Ehrenreich somewhat ironically earned a doctorate at Rockefeller University studying microphages, noting "I thought they were my friends" (Ehrenreich 2018, p. xii).

The sad realization that human bodies are innately programmed to self-destruct brought Ehrenreich to reflect on the futility of trying to outwit death through medical interventions, through lifestyle adjustments, and often expensive participation in "a nebulous but ever-growing 'wellness' industry that embraces both body and mind" (Ehrenreich 2018, p. xiv). She made a decision some years ago to forego the periodic probing that might reveal conditions requiring medical or surgical intervention. In her mid-70s she came to a place of peace with the knowledge that 1 day she will be no more. In conversation with friends who had become worried about the suffering they would most likely encounter in their final weeks, months, or years, she reassured them "that this could be minimized or eliminated by insisting on a non-medical death, without the torment of heroic interventions to prolong life by a few hours or days" (Ehrenreich 2018, p. 208).

Ehrenreich's book does not end in nihilism, but rather with a willingness to abandon attachment to self. In a profound way, she poses the question, "Who Dies?" She also recognizes a "coming deep paradigm shift from a science based on the assumption of a dead universe to one that acknowledges and seeks to understand a natural world shot through with nonhuman agency" (Ehrenreich 2018, p. 208). The microphage, rather than being her enemy, became her friend by teaching her the impossibility of any single ego-driven human to control his or her destiny.

Though Ehrenreich would reject the notion of a Jain soul in favor of the more secularly-friendly Buddhist notion of no-self, she does hold out hope for a living universe, stemming from her awe at the world of microbes and her acknowledgement of vibrations at the quantum level:

> Maybe then, our animist ancestors were on to something that we have lost sight of in the last few hundred years of rigid monotheism, science, and Enlightenment. And that is the insight that the natural world is not dead, but swarming with activity, sometimes perhaps even agency and intentionality. Even the place where you might expect to find quiet and solidity, the very heart of matter—the interior of a proton or a neutron—turns out to be animated with the ghostly flickerings of quantum fluctuations. I would not say the universe is "alive," since that might invite misleading biological analogies. But it is restless, quivering, and juddering, from its vast vacant patches to its tiniest crevices. (Ehrenreich 2018, pp. 202–203)

This description of movement in fact aligns with the Jain philosophy of *spanda* or vibration, said to lie at the heart (and soul) of life. Many people refer to Jainism as an animist faith, due to its assertion that life can be found in the vibratory patterns of earth, water, fire, and air, manifested in rocks and streams, gusts of wind and blazing fires.

Ehrenreich echoes a skepticism about the medical establishment that was similarly expressed more than a century ago by Mahatma Gandhi. In what may be deemed his quintessential manifesto, *Hind Swaraj* (1907), he vociferously protested the hegemony of medical practice. Though he wrote at a time before the advent of antibiotics, when cures were somewhat less certain, his words are prophetic, particularly in this century that anticipates the failure of antibiotics and the imminent death of 20 million persons annually from super bacteria.

The question has been asked by Ehrenreich and many others: why do people take such drastic measures to postpone the inevitable? The answer may in part be found in the Bible. From the beginning of the Christian tradition, death has considered to be an "evil," an "enemy to be destroyed" (I Corinthians 15:26). The rise of modern medicine ushered in medicines and technologies that allow the postponement of death. The marriage of faith and medical technology has created a fetish of life extension. What once was known as inevitable now seems not only unfair but also evil, if not sinful.

7.9 Gandhi, Medicine, and Death

For Gandhi, death was not a mystery, nor were extraordinary measures available widely to avoid death during his lifetime, at least not to the degree possible today. Because he grew up with many Jain friends, the tradition of fasting unto death undoubtedly informed his worldview. Death with dignity would have been part of Gandhi's experience. He considered the manner of death far more important than the avoidance of death. He advocated a spiritual approach to death. He considered medicine to be a "parasitical profession" (Parel 1997, p. 62), and he wrote, "Doctors have nearly unhinged us" (Parel 1997, p. 63). He held a highly suspicious attitude toward medicine, and he regarded reliance on physicians as feeding human weakness:

> How do these diseases arise? Surely by our negligence and indulgence. I over-eat. I have indigestion, I go to a doctor, he gives me medicine, I am cured, I over-eat again, and I take his pills again. Had I not taken the pills in the first instance, I would have suffered the punishment deserved by me, and I would not have over-eaten again. The doctor intervened and helped me to indulge myself. My body thereby certainly felt more at ease, but my mind became weakened. A continuance of a course of medicine must, therefore, result in loss of control over the mind.... Had the doctor not intervened, nature would have done its work, and I would have acquired mastery over myself, would have been freed from vice, and would have become happy. (Parel 1997, p. 63)

Though this approach seems to arrogate all illness to the mind without taking into account truly debilitating conditions beyond human control, Gandhi nonetheless offers a measure of insight and wisdom. The rise of diabetes worldwide is due in part to the increase of caloric intake. Many diseases, including those related to alcoholism and smoking, arise due to human behavior.

Anticipating the rapaciousness of pharmaceutical companies, he also observed:

> Doctors make a show of their knowledge, and charge exorbitant fees. Their preparations, which are intrinsically worth a few pennies, cost shillings. The populace in its credulity and in the hope of ridding itself of some disease, allows itself to be cheated. (Parel 1997, p. 65)

As one reflects on the disproportionate compensation given to physicians, particularly in the United States, and the extreme expense of both medicine and insurance, Gandhi's homespun remarks continue to be poignant.

Gandhi proclaimed that a doctor, in order to exert the strength needed for true independence or *swaraj*, "will give up medicine, and understand that, rather than mending bodies, he should mend souls" (Parel 1997, p. 117). In a radical acceptance of the inevitable, he considered it to be nobler for a patient to succumb rather than subject himself or herself to drugs:

> if any patients come to him, [the doctor] will tell them the cause of their diseases, and will advise them to remove the cause rather than pamper them by giving useless drugs; he will understand that, if by not taking drugs, perchance the patient dies, the world will not come to grief, and that he will have been really merciful to him. (Parel 1997, p. 117)

Although this advice would often be imprudent today given the advances of modern medicine, the adoption of a more non-interventionist approach would help reduce the modern doctor's propensity for ordering expensive tests and prescribing too many medications. Ehrenreich would agree wholeheartedly with this assessment.

Gandhi advocated simple living. He held deep suspicions about the benefits of European civilization, including its insistence on using medicines of dubious efficacy. The wisdom he shared regarding self-reliance remains relevant for the contemporary post-modern world, though with modifications. To use the contemporary examples of the deaths of Mrs. Vijay Bhade and John Rehm as well as the decision by Barbara Ehrenreich to forego routine testing, the calmer approach to the inevitability of death would be to find acceptance rather than seeking yet another procedure or medicine.

7.10 Conclusion

In this chapter we have explored the Jain worldview that, because of its undying faith in the eternality of the soul, does not seek to avoid death. Jains are encouraged to embrace death when all reasonable options have been explored. The practice of fasting to death, though under legal attack in India, holds some instructive wisdom and in fact has been an important option for such persons in America as Scott Nearing, Mrs. Bhade, and John Rehm.

The advent of hospice care and the enactment of right-to-die laws signal a new moment in the western approach to death and dying. When the California law authorizing physician-assisted suicide passed in 2015, more than 100 people chose this option within 6 months (Karlamangla 2018). One in five Americans live in a state that allows physician-assisted suicide. However, similar to the resistance against the Jain final fast in India, the California law allowing physician-assisted suicide has been challenged in court and might be overturned. Pro-life forces are strong on both continents.

It is important not to conflate physician-assisted suicide with fasting-to-death. Medically induced death would not be allowed within the Jain faith. The physicians administering the "medicine" would accrue burdensome karma for performing what Jains consider a violent misdeed. The dying patient would be regarded as not possessing the needed courage to give up attachment to the body. Nonetheless, both the fast unto death and advocacy for physician-assisted death arise from a concern that it is not prudent or wise to prolong life at any cost.

By recognizing that the human condition is fraught with frailty and that all treatments to prevent death will eventually fail, an acceptance of death can arise. From this acceptance, decisions can be made from a place of reason, not from a place of emotional attachment or clinging to a life that cannot be made viable. Discerning the right course of action can be very difficult, particularly in this era of medical advancements. One must weigh the benefits and risks of any medical treatment carefully, and in consultation with family and friends. However, rather than running blindly away from death, it can be better to reckon with its looming inevitability, in the words of Lucy Bregman: "To befriend death is to open oneself up to a sweeping, total encounter with its reality as hidden presence in one's ongoing life" (Bregman 1992, p. 214). To die a resistant death doubtlessly increases suffering. To die in a mental and emotional state that accepts death, though perhaps not any less physically uncomfortable, can bring peace.

References

Babb, Lawrence A. 1996. *Absent lord: Ascetics and kings in a Jain ritual culture*. Berkeley: University of California Press.

Braun, Whitny. 2015. *Sallekhana: The philosophy, ethics, and culture of the Jain end of life ritual*. Doctoral Dissertation, Claremont Graduate University.

Bregman, Lucy. 1992. *Death in the midst of life: Perspectives on death from Christianity and depth psychology*. Grand Rapids: Baker Book House.

———, ed. 2010. *Religion, death, and dying*. Vol. 3. Santa Barbara: Paeger.

Caillat, Collete. 1977. Fasting unto death according to Jaina tradition. *Acta Orientalia* 38: 43–66.

Chapple, Christopher Key. 1993. *Nonviolence to animals, earth, and self in Asian traditions*. Albany: State University of New York Press.

———. 2010. Eternal life, death, and dying in Jainism. In *Religion, death, and dying, volume I: Perspectives on dying and death*, ed. Lucy Bregman, 198–212. Santa Barbara: Paeger.

deNicolas, Antonio T. 2004. *The Bhagavad Gītā: The ethics of decision-making*. Berwick: Nicolas-Hays.

Ehrenreich, Barbara. 2018. *Natural causes: An epidemic of wellness, the certainty of dying, and killing ourselves to live longer*. New York: Twelve.

Ehrenreich, Barbara, and Deidre English. 1978. *For her own good: Two centuries of experts' advice to women*. New York: Anchor Press.

Jaini, Padmanabh S. 1979. *The Jaina path of purification*. Berkeley: University of California Press.

Karlamangla, Soumya. 2018. State fights to uphold its assisted suicide law. *Los Angeles Times*, May 22, 2018.
LaConte, Ellen. 1997. *Free radical: A reconsideration of the good death of Scott Nearing*. Stockton Springs: Loose Leaf Press.
Nearing, Helen, ed. 1995. *Light on aging and dying*. Gardiner: Tilbury House.
Nuland, Sherwin B. 1994. *How we die: Reflections on life's final chapter*. New York: Alfred A. Knopf.
Parel, Anthony J., ed. 1997. [1907]. *Hind Swaraj and other writings by M. K. Gandhi*. New York: Cambridge University Press.
Rehm, Diane. 2016. *On my own*. New York: Alfred A. Knopf.
Settar, Shadakshari. 1986. *Inviting death: Historical experiments on Sepulchral hill*. Dharwad: Institute of Indian Art History.
———. 1990. *Pursuing death: Philosophy and practice of voluntary termination of life*. Dharwad: Institute of Indian Art History.
Tukol, T.K. 1976. *Sallekhana is not suicide*. Ahmedabad: L.D: Institute of Indology.

Chapter 8
The Ritualization of Death and Dying: The Journey from the Living Living to the Living Dead in African Religions

Herbert Moyo

Abstract This essay examines African understandings of life, sickness, death, and life after death by concentrating on the Ndebele people of Matabo in Zimbabwe who are part of the Nguni people of Southern Africa and who have a strong Zulu cultural basis. For the Ndebele of Matabo, dying is a physical, medical, and spiritual phenomenon. Healing through medicine is always prioritized when there is a sickness, but all medical treatment is subordinate to the spiritual world. The Ndebele understand that when people eventually die, despite medical attention, it is a sign that the spiritual world is more powerful than the medical world. Those who die are said to have responded to a call from their ancestors. For them, death is a transition from the world of the living living to the world of the living dead. This essay will briefly explore the African Ndebele concepts of life, death, the ritualization and medicalization of death and dying, and life after death. It will argue that, for the Ndebele, death is not a medical phenomenon but a response to a calling by ancestors to the spiritual world at the fulfilment of one's time on earth as determined by the *abaphansi*. Death, therefore, is ritualized, not medicalized.

8.1 Introduction

In the Ndebele worldview, death does not end life but continues it in the realm of the ancestors. Death does not affect the personality of the dead person but only changes the living conditions; thus, the dead are merely the "living dead." (Harriet 1977; Andersen n.d.). Indeed, the ancestors (*abaphansi*) continue to live and interact with the living living. This concurs with Setiloane's observation that the African understanding of community includes the extended family of the living living and the living dead (Setiloane 1986). When a person had died, "the Zulu would say *useye*

H. Moyo (✉)
School of Religion, Philosophy and Classics, University of KwaZulu-Natal, Pietermaritzburg, South Africa
e-mail: Moyoh@ukzn.ac.za

T. D. Knepper et al. (eds.), *Death and Dying*, Comparative Philosophy of Religion 2, https://doi.org/10.1007/978-3-030-19300-3_8

koyise-mkhulu (he/she has gone to be with the forefathers); when a Zulu says the praise—songs of the living dead—he would end by saying *asifi siyalala* (we do not die but sleep)" (Ngobese 2009).

There is an inherent belief among the Ndebele that the journey to the world of the living dead has many obstacles and interruptions; hence the insistence on the observation of the correct funeral rites. If the correct funeral rites are not observed, the deceased may come back to trouble the living relatives or may refuse to come back to look after the family as an ancestor.

The Ndebele culture has been corrupted by other traditions and cultures as a result of globalization. The major agent for westernization of the Ndebele was and continues to be the Christian church, through both western missionaries and contemporary black African pastors who have been indoctrinated with Western systems of knowledge. Westernization of the Ndebele in particular and many parts of Africa in general cannot therefore be separated from the influence of Christianity, which is synonymous with the teachings and practices of historically missionary churches (e.g., Lutherans, Seventh Day Adventists, Catholics, Anglicans, Presbyterians, and the Church of Christ). In Matabo, all the hospitals and high schools belong to the Lutheran church.[1] The tensions between African Indigenous Religion(s) (AIR) and medicalization are therefore tensions between AIR and Christianity, as played out in medical institutions and on occasions of dying and death.[2]

The socio-religious space is a negotiated space with several dynamics that manifest themselves through hostility, demonization, tolerance, and inculturation, as each situation arises. These can be pronounced where the family of the dying or dead person belongs to different religions (or denominations) and different members claim authority over the rituals to be performed—for example, if siblings do not agree on the rituals to be performed for a dying parent or some other relative who is recently deceased. But the people of Matabo are also very accommodating to practices from a number of religions that are performed for a dying or dead person. Suffice it to say, there are dynamics in the relationship between AIR, Christianity, and the Westernization of death and dying rituals in Matabo that show high levels of syncretism (Moyo 2014).

Despite the influence of Western civilization and Christianity, however, the living dead remain active in the community and are consulted on significant occasions to intervene on behalf of the living living. The living dead are accorded great respect and even fear in some instances (Schapera 1937; Nxumalo 1966; Msomi 2008).

This paper begins with a brief overview of who the Ndebele people are. It then examines the Ndebele religious and cultural beliefs around death and dying as well as relationships with the ancestors. It goes on to discuss the tensions between traditional views of death and dying and the medicalization of death and dying, as well

---

[1] Matabo is part of the greater Mberengwa district of Zimbabwe. It is uniquely Ndebele-speaking in a Shona-speaking province.

[2] AIR is commonly referred to as African Traditional Religion (ATR). I prefer AIR over ATR, since I see Christianity as an ATR but not an AIR due to the fact that it is now more than three centuries old in Africa. How long does it take for a practice to become a tradition?

as the challenges of dying in the hospital (away from home). This paper then looks at the family graveyard as the ideal burial place and burial rituals such as the cleansing of the people present at the burial and the tools used during the burial. Finally, it turns to the spirit bringing-home ceremony (*umbuyiso*).

8.2 Who Are the Ndebele People of Matabo?

The Ndebele of Zimbabwe are Zulus who travelled northwards from South Africa in the nineteenth century. In 1823, Chief Mzilikazi Khumalo had a conflict with King Shaka Zulu over the spoils of war in KwaZulu-Natal. As a result, Mzilikazi travelled north with his regiment and their families. Eventually Mzilikazi settled in Zimbabwe, becoming the king of the Ndebele. The name "Ndebele" comes from the Sotho people who called them "Matabele," which means "people of the shields," in reference to the shields of war that these Zulus carried.[3] As these Zulus travelled northwards, their vocabulary became richer (or polluted) by picking words from the tribes with which they interacted. So the Ndebele of Zimbabwe speak what some people would call corrupted Zulu.

Matabo is an area in the Midlands province bordering Matabeleland South Province. In present day Matabo, there is virtually no home or family without someone living and working in South Africa. This has created cultural links between Matabo and the Zulu in South Africa, especially for the Ndebele, who see this connection as a link to their origin. These links offer interesting cultural synergies; hence my focus on the cultural practices of the Ndebele of Zimbabwe (not the Ndebele of South Africa) and their relationship with the Zulu of South Africa, which whom they share the same language group, cultural observances, and religious practices.

8.3 The Religion and Culture of the Ndebele

The Ndebele are very religious people who find it difficult to make a distinction between their Ndebele culture and AIR; for them, there is no life outside the realm of religion. African Indigenous Religion (AIR), which permeates life from the womb and continues into life after death, is the total traditional worldview with all the rituals, principles, values, and beliefs. In most cases religion means the presence and the influence of God (the creator of the Ndebele nation),[4] ancestors, and the

---

[3] This paper presents only a summarised version of the history of the Ndebele people.

[4] The Ndebele culture is in conversation with Christianity. The Christian God is commonly written with a capital letter "G" whereas any other God is written with a small letter "g." This has the connotation of subordinating other gods to the Christian God. But the Ndebele God is God (not god) for the Ndebele. The use of capital letters and small letters has political connotations among the

spirit world generally. Bozongwana rightly says that African religion does not need to convert or proselytize, since one is born into it (Bozongwana 1983). In agreement with Bozongwana, I maintain that African religion does not wait for a person to grow up to learn how to read about the religion from some kind of "Holy Book"; its strength is that it is a lived religion from birth to death and on to the world of the ancestors (Moyo 2014).

The Ndebele cosmology includes both visible and invisible spirit beings that influence living human beings and are appeased by using rituals and living life according to acceptable social norms and values. (Moyo 2014, p. 117). Similarly, Mbiti notes that for the Ndebele, "the unborn, the living living and the living dead are all part of the constellation of the living spirits" (Mbiti 1989). Ndeti too says that in the African community "(the individual) is a physical representative of the dead, living and unborn. Thus (the individual) is a community incorporating three principles—life, spirit and immortality" (Ndeti 1972, p. 114). In agreement, Masamba says that the ancestors have special powers that "enable the birth of children and protect the living family members from attack by malevolent spirits" (ma Mpolo and Nwachuku 1991, p. 24). Given the ancestors' ability to punish, exonerate, and reward, "the health of the living depends to a great extent on their relationships within the extended family and with their ancestors" (ma Mpolo and Nwachuku 1991, p. 24).

Life amongst the Ndebele is firmly rooted in this cosmology and worldview, making it very difficult to separate culture from religion. When things do not work out in life, one is forced to seek an understanding of their relationship with the living dead and God. This understanding of God and ancestors borders around a theology of retribution. God and the ancestors are happy, and they bless those who relate well to others and the environment. On the contrary, those who do not relate well will be cursed through misfortune and ill-health. The ancestors can only be consulted if the dead have had all the necessary rituals performed accordingly. Thus Ngobese says,

> This comes out clearly in burial rituals and farewell speeches of the dying Africans.... The following [is a typical funeral prayer from] the Zulus: "Farewell, our father. Protect us your children from suffering, starvation, disease, death of stock, drought and from barrenness." (Ngobese 2009, p. 25)

8.4 Tensions with the Medicalization of Death and Dying

The Ndebele are well aware of the benefits of the medicalization of diseases and sicknesses, specifically its potential "to improve the experience and quality of life of the affected individual" (Gitome et al. 2014, p. 25). It is for this reason that the

---

Ndebele who feel that Christianity is belittling their (Ndebele) God. So from the perspective of a post-colonial or de-colonial theory, one should make an effort not to subordinate the Ndebele God(s) to the Christian God.

Ndebele are willing to send their sick to hospitals for medical attention. However this understanding of the medicalization of diseases is located within a traditional Ndebele view that death is the result of a calling by the ancestors, not disease or sickness. Diseases can be cured by hospitals and local herbalists only when ancestors still allow one to live on earth. This belief that death is a calling to the land of the living dead necessitates elaborate rituals that must be performed correctly to meet the needs of the ancestors.

8.5 Challenges of the Hospitalisation System

8.5.1 Communal Visiting of the Sick in Hospitals

The Ndebele strongly prefer that their beloved die at home surrounded by family, friends, neighbors, and relatives. They cannot leave a sick person "alone" with strangers; thus they will go as a clan to visit a sick relative in the hospital. Those belonging to the Church will pay a visit to the hospital as a congregation. In the traditional worldview of the Ndebele, those who die alone in the hospital surrounded by strangers are not only socially isolated but also robbed of the opportunity to tell their family how to perform rituals for them (Seymour 2001). For traditional Ndebele, hospital death is therefore dehumanized dying (a fact that is exacerbated by the failure of Western medical systems to coexist appropriately with African traditional medicine and ritual).

Local hospitals such as the clinics at Masase, Mnene, Msume, Gwanda, and Kotokwe have legislated that only two visitors per patient will be allowed, and they provide security to enforce this regulation. This too is a challenge, since it goes against the communal nature of the life of the Ndebele. In most cases, the sick want to be visited by as many people as possible. Moreover, the family and friends of the sick must visit as a form of social responsibility. In fact, the whole village of the sick person would come to visit if not for financial constraints. This goes against the "the two visitors per patient rule," which seemingly was created for black Africans.

8.5.2 The Challenge of Leaving the Patient with Strangers

Relatedly, a strong Ndebele ethic of care for the sick family members survives despite the advent of hospitals. When my father was hospitalised, my uncle, an elder brother to my father, demanded permission to sleep next to his sick brother in the hospital ward. Naturally, the hospital refused. He became so stubborn that the hospital manager eventually gave him bedding to sleep next to his brother. The Ndebele struggle with leaving their relatives in the care of strangers. In the majority of cases, the hospital does not permit relatives of patients to stay with them; thus they must find accommodations near the hospital.

8.5.3 Collecting the Spirit of the Deceased from the Hospital

If a family member dies away from home, many Zulu and Ndebele families believe the family must collect the spirit of the deceased from the exact place of death. This is an elaborate exercise where a family representative goes to that place to speak to the spirit of the dead person, telling him or her that they are now going home. Usually the oldest person will speak words such as:

> X, we are now taking you home, we are leaving place Y to the home of your forefathers in Matabo koBankwe. Please agree to go with us, do you hear us? Please make the journey easy, do not create problems along the way. We are using a car so help us to avoid accidents and breakdowns along the way. It is me Z and family members A, B, and C. Let us go.

In some cases these instructions are followed by the praise names that constitute the genealogy and history of the exploits of the family of the deceased. This can only be done by a biological family member; otherwise the deceased will not respond accordingly. Next, the spirit will be informed about the route to be travelled, such as rivers to be crossed and accidents that have happened. And once the party begins this route, they must be prepared to stop along the way to allow the deceased time to rest. Finally, when the party arrives at the home of the deceased, the spokesperson will announce, "We have now arrived at the home of your forefathers. We are now moving you out of the car into your house. Thank you for having enabled us to have a safe trip from the mortuary to our home."

This ritual can be problematic if the hospital takes the body of the deceased to a mortuary and gives the bed to another patient. The family of the deceased will still go to the death bed to talk to the spirit of their deceased. Imagine the trauma for the next patient who inherits that bed.

8.5.4 Collecting the Spirit of a Person Who Dies in a Moving Ambulance

Another dilemma occurs if the person dies inside a moving ambulance (or any other moving vehicle). If the ambulance does not stop at the point of death but rather proceeds to the hospital or mortuary, it becomes difficult to know from where to collect the spirit of the dead person. There are instances where families try to identify the actual ambulance and collect the spirit from the vehicle, while others simply try to collect the spirit from the mortuary. Both violate the proper way of collecting the spirit of the deceased, which is from the exact location of their death. Spirits that are not collected properly and taken home will not look after the family as an ancestor.

8.6 Other Aspects of the Medicalization of Disease

As a hospital chaplain, I have observed that the Ndebele are keen to try African traditional medicine that comes from traditional healers, who in most cases are convincing since they have consulted ancestors about the best possible treatment. Relatives then try to smuggle traditional remedies to the hospital. This always creates conflict between Western-trained doctors and traditional Ndebele doctors.

There is also the challenge of blood transfusions and organ transplants. For the Ndebele, blood relationships are based on the purity of the blood of the relative. When they say *leli ligazi lami* ("this is my blood"), they mean that someone belongs to their genetic lineage. So if you get blood from someone who is not your blood relative, you dilute the relationship.

Similarly, the heart is an organ that pumps blood, but for the Ndebele it is what is used to think, make decisions, and carry the emotions of the person. If you give a person a heart from another person, then that person will belong to two families, which is unheard of in the Ndebele culture. Even worse is what the Ndebele imagine will happen when the person with someone else's blood or heart dies: ancestors may not accept this person in the world of the living dead, which is a disaster.

Some traditionalists among the Ndebele would rather die than be given blood or an organ that can hinder their eventual transition to the world of the ancestors. In both cases, the hospital is seen as having the potential to pollute the person to the point of rejection by the ancestors.

8.7 The Family Graveyard as the Ideal Burial Place and Related Rituals

Even though the emphasis is on the spirit of the deceased, the body of the deceased is also very important, as the spirit must be in close proximity to its dead body. Those who die in hospital must be transported home. Even though many now work in foreign countries, the Ndebele still expect their deceased to be transported back to Zimbabwe.

Ideally, the deceased are buried in the family graveyard next to their forefathers. This is the first physical link to the ancestors (besides the announcements that were made as he/she is carried to the grave). The body must be laid inside the grave either on their right side or left side and never facing downwards or upwards inside the grave.[5] This can be difficult when coffins are used, since coffins are sometimes too small to be closed if the dead body is laid on its side.

[5] Which side it is laid on depends on where it is buried with respect to forefathers—the deceased must face in the direction of those who died before them (Moyo 2014, p. 119).

When the Ndebele talk to their ancestors, they must go to their grave. For this reason, the possibility of cremation of the body is not even discussed among the Ndebele of Matabo. This also brings about tension with the Church in cases where Christian missionaries developed communal gravesites next to church buildings, thus taking the deceased away from the family graveyard.

When the Ndebele talk to deceased family members, they ask for the living dead to look after the family they are leaving behind and, ultimately, to help them find their way to the world of the living dead upon death (Berglund 1976).[6] This can be asked during a Christian funeral, especially when there is viewing of the deceased[7]; alternatively, it can be asked when people are laying flowers on the tomb after covering it with soil. Each flower comes with a written message. Usually the message is directed towards the dead person with messages such as *hamba kahle sizokukhumbula sibanibani* ("go well we will miss you/remember you"), *hamba kahle usilungisele indawo lathi sizakulandela* ("go well and prepare a place for us where you are going, we will follow you"), *ungakhohlwa abantwana obatshiye ngemuva* ("do not forget the children that you left behind"), *phumula kahle abantwana bakho sizabagcina* ("rest in peace, we will look after your children") (Mapumulo 2013, p. 5). These messages are a sign of a strong belief in life after death and the power that is assumed by those who have died. People start voicing these petitions "while the person is still fresh in the grave" (Moyo 2014, p. 119).

8.8 The Cleansing of the People and Tools Used During the Burial

Within a month after the funeral, the family of the deceased organizes a ceremony for the cleansing of the tools used during the funeral and sometimes also the cleansing of people who participated in the funeral. This is necessary to prevent the death from being carried to the family members of those involved in the burial. Traditionally, it is important enough to warrant its own separate ritual (Sitshebo 2000, p. 51).

During the tool-cleansing ceremony, a family representative who is believed to be strong enough to address those that are gathered for the ceremony (usually the oldest man in the family of the deceased) addresses the gathering, thanking people for having helped in the funeral. This family representative then discloses the cause of death to the gathering. In most cases there will be suspicions of witchcraft. If the

---

[6] For Christians the world of the living dead is heaven, whereas for adherents of AIR it is *phansi* (down). There are some Ndebele who speak of both worlds to their deceased.

[7] The primary agent for westernisation in Matabo in particular and Southern Africa in general is the Church. The gospel and western culture came to Africa as one package. The contemporary cultural practices around death and dying in Matabo are always in conversation with Christianity. In Matabo all funerals, whether for Christians or members of AIR, are conducted by a Christian minster. This is a new practice and culture for Matabo.

deceased was murdered or died mysteriously (e.g., if he/she was found dead), this family representative will incite the dead person not to sleep and relax but to fight back and strike the killers (Moyo 2014, p. 120–121). Words such as *vuka uzilwele, ungalali ubuthongo* ("wake up and fight, do not sleep") are spoken (Banana 1991, p. 83). After this address, those present drink homemade beer and eat meat that has been dedicated to the ancestors.[8] Then they are free to collect their tools that have been used in the burial, which no longer have dirt on them and are no longer contaminated by death (Sitshebo 2000, p. 52; Bozongwana 1983, p. 29).

There are a few instances where you see this ceremony being conducted by the Church through prayers and mass. Usually, however, this ceremony is held independently of a Christian service. In either case, the Ndebele comingle AIR and Christianity (Moyo 2014, p. 120).

8.9 *Umbuyiso*: The Spirit Bringing-Home Ceremony

The tool-cleansing ritual is followed by the calling of the spirit of the dead person to come home and look after the family. This is called *umbuyiso*. Since about 1992, Christianity has created its own ceremony called *ukwambulwa kwetshe lethuna* (tombstone unveiling) or *isikhumbuzo* (memorial). What is interesting to observe is that *umbuyiso* is done in the case of the death of a respectable family member who had a wife and children (Daneel 1987, p. 237). The Christian ceremony is different from the *umbuyiso* in that it is done for all deceased people regardless of their social status. Some families combine both ceremonies, performing *umbuyiso* on Thursday and Friday with traditional spirit mediums and healers, then following this with a tombstone unveiling on Saturday conducted by a church minister.

The *umbuyiso* symbolizes the spiritual resurrection of the dead, appeasing the spirit of the dead by calling it back home, thus recognizing that the deceased still lives in spiritual form (Moyo 2014, p. 121). The *umbuyiso* is therefore one meeting point of Christianity and AIR, with both confessing life after death (even though the nature of such life may not be the same in both cases). As noted earlier, "death does not sever the bond between the living and the dead" for the Ndebele (Dickson 1984, p. 193).

8.10 Conclusion

This paper concludes that the Ndebele ritualize death and dying in a quest to create harmonious relationships with the living dead, thereby at times creating tensions with the medicalization of death. Despite the increased medicalization of diseases,

[8] Usually a cow or a goat is prepared, depending on the economic status of the family of the deceased (Moyo 2014, p. 119).

the Ndebele continue to find ways of practicing their cultural beliefs, negotiating the border between the western medicalization and traditional ritualization of death and dying. Death is not medicalized but ritualized.

References

Anderson, Allan. n.d. African religions. *Encyclopedia of death and dying*. http://www.deathreference.com/A-Bi/African-Religions.html#ixzz4J4fHPCDj. Accessed 08 Aug 2016.

Banana, C.S. 1991. *Come and share: An introduction to Christian theology*. Uppsala: Swedish Institute of Missionary Research.

Berglund, A. 1976. *Zulu thought patterns and symbolism*. London: Hurst.

Bozongwana, W. 1983. *Ndebele religion customs*. Gweru: Mambo Press.

Daneel, I. 1987. *Quest for belonging*. Gweru: Mambo Press.

Dickson, K.A. 1984. *Theology in Africa*. London: Darton Longman & Todd Ltd.

Gitome, Serah, Stella Njuguna, Zachary Kwena, Everlyne Ombati, Betty Njoroge, and Elizabeth A. Bukusi. 2014. Perspectives paper: Medicalization of HIV and the African response. *African Journal of Reproductive Health* 18: 25–33.

Harriet, Ngubane. 1977. *Body and mind in Zulu medicine: An anthropology of health and disease in Nyuswa-Zulu thought and practice*. London: Academic.

Mapumulo, Z. 2013. *Mido Macia goes home*. City Press, 10 March, p. 5.

ma Mpolo, Jean Masamba, and Daisy Nwachuku. 1991. *Pastoral care and counselling in Africa today*. New York: Peter Lang.

Mbiti, J.S. 1989. *African religions and philosophy*. Westlands: East African Educational Publishers Ltd.

Moyo, H. 2014. Dual observances of African traditional religion and Christianity: Implications for pastoral care in the pluralistic religious worldview of the Ndebele people of Matabo in Zimbabwe. *Journal of Theology for Southern Africa* 148: 115–132.

Msomi, V.V. 2008. *Ubuntu contextual African pastoral care and counselling*. Pretoria: CB Powell Bible Centre-UNISA.

Ndeti, K. 1972. *Elements of Akamba life*. Nairobi: East African Publishing House.

Ngobese, Wilmot Ronald Musa. 2009. *The continuity of life in African religion with reference to marriage and death among the Zulu people*. Pretoria: University of South Africa.

Nxumalo, O. 1966. *Inqolobane Yesizwe: A survey of Zulu culture and customs*. Pietermaritzburg: Shuter and Shooter.

Schapera, I. 1937. *The Bantu speaking tribes of South Africa*. London: Lund Humphr.

Setiloane, G.M. 1986. *African theology: An introduction*. Braamfontein: Skottaville Publishers.

Seymour, J.E. 2001. *Critical moments: Death and dying in intensive care*. Buckingham: Open University Press.

Sitshebo, T. Wilson. 2000. *Towards a theological synthesis of Christian and Shona views of death and the dead: Implications for pastoral care in the Anglican diocese of Harare, Zimbabwe*. Birmingham: The University of Birmingham Press.

Chapter 9
Death in Tibetan Buddhism

Alyson Prude

Abstract Tibetan Buddhist understandings of the death process bridge scientific, materialist observations and religio-spiritual interpretations. Tibetan Buddhism and medicine overlap in the context of death, and doctors of Tibetan medicine are trained in tantric Buddhist theories as well as anatomy and pharmacology. Based on written as well as contemporary oral sources, this essay explores Buddhist concepts of the process of dying and the experiences of consciousness as it transitions from one life to the next. It describes end-of-life rituals and funerary customs, notions of timely versus untimely death, and the possibility of returning from death to one's previous life. A Tibetan Buddhist perspective challenges reductionist western medical theories that refuse to allow for the continuation of consciousness at death. In modern medical contexts, Buddhist theories of karma and rebirth influence treatment decisions such that, as opposed to prolonging life for as long as possible, people are encouraged to accept the certainty of death and learn to face death with less anxiety. From a Tibetan Buddhist perspective, a good death is more valuable than a prolonged life.

9.1 Introduction

According to a Buddhist perspective, death is not often a permanent end to existence. Instead, death is merely the separation of consciousness from the physical body, and dying marks the beginning of rebirth into another life. Like many religious practitioners who believe in the continuation of life after death, it is common to hear Tibetan Buddhists claim that they are not afraid to die. For them, a person's death marks the end of that particular human birth and the start of a new life in a new body, perhaps human but more likely of another sort. Tibetan Buddhists refer to cases of children speaking of past lives, the experience of déjà-vu, the phenomenon of people who return from death, and the bodies of great meditators that do not

A. Prude (✉)
Georgia Southern University, Statesboro, GA, USA
e-mail: mprude@georgiasouthern.edu

T. D. Knepper et al. (eds.), *Death and Dying*, Comparative Philosophy of Religion 2, https://doi.org/10.1007/978-3-030-19300-3_9

immediately decompose at death as positive empirical evidence substantiating Buddhist claims. A Tibetan Buddhist view of death thus challenges reductionist western medical theories that refuse to allow for the continuation of consciousness at death. In fact, if death is defined as the cessation of consciousness, then Tibetan philosophy refutes death.

As regards a Tibetan Buddhist view of dying, one could offer a philosophical definition or a list of physiological indications. Either way, the lists would not be wholly distinct; each would have direct correlates with the other. The overlap between philosophy and physiology is a result of the intermixing of "religious" and "medical" theories and procedures in Tibetan texts. The *Tengyur* (*bsTan 'byur*), the accepted commentaries on the Buddha's teachings, for example, includes Indian Ayurvedic treatises. Likewise, the *Kalachakra Tantra* and the *Four [Medical] Tantras* (*rGyud bzhi*), the latter serving as the foundational text for training in Tibetan medicine (*gso ba rig pa*), present philosophy, psychology, soteriology, and pharmacology alongside one another.[1] As Wallace has shown, medical theories and practices hold important "soteriological significance… on the Buddhist Tantric path" (Wallace 1995, p. 155). Behind the combination of spirituality and medicine lies the tantric view that one's physical and spiritual health are mutually influential, and spiritual liberation is achieved in dependence upon a properly-functioning body.

As a result, Tibetan medical and religious literature intersect, and religion and medicine do not exist in opposition.[2] The main training center for Tibetan medicine, the Men-Tsee-Khang (sMan rtsis khang) or Tibetan Medical and Astro-science Institute, counts among its many services the provision of almanacs and horoscopes, the distribution of medicinal herbs, and the publication of medical and astrological treatises, as well as western-style clinical research on diseases such as cancer and diabetes.[3] Buddhist monks and teachers played a crucial role throughout the history of Tibetan medicine, and today traditional Tibetan doctors (*a mchi*) may also be Buddhist monastics.

This essay investigates the overlap between Tibetan Buddhism and medicine in the context of death. It explores Buddhist concepts of the process of dying and being reborn, end-of-life rituals and funerary customs, notions of timely versus untimely death, the possibility of returning from death to one's previous life, and ways the theories of karma and rebirth play out in modern medical contexts. The content that follows is based on both written and contemporary oral sources.

---

[1] For a brief history of the *Four [Medical] Tantras*, see Fenner (1996).

[2] For examples and history of the interrelation of Tibetan medicine and religion, see Garrett (2009) and Gyatso (2004).

[3] More information about the Men-Tsee-Khang can be found on their website, http://men-tsee-khang.org/index2.htm, which features images of the Medicine Buddha and the Dalai Lama on its homepage.

9.2 The Cycle of Rebirths

Tibetan Buddhist ideas about the relationship between consciousness and the body are based on an Indic tantric model according to which consciousness is a sort of vital principle that rides on or is carried by the body's energies or winds (*rlung*, Skt. *prāṇa*) as they circulate through the veins or channels (*rtsa*, Skt. *nāḍī*). A common metaphor for describing the relationship between the consciousness and the body is that of people living in a house. The house, in this example, is the body, a physical container for the people who represent the lively, activity-producing consciousness. Consciousness is thus seen as intimately connected to yet ultimately distinct from the physical body. Just as people can leave a house and go outdoors, consciousness is believed to be able to exit the body. This is what happens at death. Tulku Thondup, a Tibetan Buddhist scholar, writes, "When breathing stops and the [body's] heat is lost, the mind and body separate, and the person is dead" (2005, p. 17). Tibetans with whom I spoke defined death in similar terms. "When consciousness gathers together and rids itself of the body, this is called dying," an educated layman told me.

Dying is followed by rebirth. Tibetan Buddhist teachings specify six classes of birth in the desire realm of *saṃsāra* (the cycle of death and rebirth). Which realm a being takes rebirth into is largely determined by that being's karma, the effects of past thoughts and actions. Buddhists posit a correspondence between specific types of thoughts and actions and particular types of rebirth. Generosity and goodwill lead to rebirth as a god, a form in which one enjoys an exceedingly long life of pleasure and the ability to intercede in the human world, if one so desires. Depicted on the Wheel of Life (*srid pa'i 'khor lo*, Skt. *bhavacakra*) as bordering the god realm is the land of the demi-gods or titans (*lha ma yin*; Skt. *asura*). Existence as a demi-god is characterized by jealousy and competitiveness, the same traits that lead to that rebirth. Next in the hierarchy is the human realm, the most desirable place to be reborn, because human life offers the best opportunities for achieving *nirvāṇa*, liberation from *saṃsāra*. With a fortunate human birth come the mental and physical abilities that make the pursuit of *nirvāṇa* possible, and the mixture of happiness and suffering that humans experience provides optimal motivation for embarking and persevering on a spiritual path. This is not the case for the fourth class of beings: animals. Animals, a category that includes insects and single-celled organisms, are said to be limited by a shortage of intellect, governed solely by instinct, and subject to control by humans. Rebirth as an animal is believed to be the result of laziness and willful ignorance and is considered an unfortunate rebirth. Another undesirable rebirth is rebirth as a ghost. Ghosts may be of many types, the most noteworthy being a "hungry ghost" (*yi dvags*, Skt. *preta*). Hungry ghosts are so called because they have huge bellies while their necks are as thin as needles. Their inability to satisfy their hunger and thirst results in a constant state of desire and craving, the same states of mind that lead to birth in that form. Finally, the hell realm is the destination of beings who have engaged in hatred and violence. The numerous hot and cold hells, like much of Tibetan Buddhist cosmology, are drawn from Indian sources and include the worst and lowest hell, Avīci.

The only alternative to rebirth, according to Buddhists, is complete and final liberation from *saṃsāra*. Practitioners who have accumulated the storehouse of merit and wisdom needed to achieve liberation may be born in paradisiacal realms, called Pure Lands (*zhing khams*), where they complete their journey to enlightenment without danger of falling back into *saṃsāra*. Tibetan and Mahāyāna Buddhists may pray to a particular buddha or bodhisattva to intercede on their behalf and secure their rebirth in that buddha or bodhisattva's Pure Land from which one is guaranteed liberation.

For beings who remained trapped in *saṃsāra*, death means another life and the continued accumulation of karma. One's karma does not ripen immediately, nor does all the karma from one lifetime manifest in a subsequent life. Instead, one's collection of karmic seeds extends indefinitely into the past, and only an enlightened being can know when the effects of a particular deed will be felt. For this reason, ordinary people cannot know the rebirth that they will take, and they may seek the advice of lamas (*bla ma*) and other Buddhist teachers for guidance regarding both their and others' future lives. Birth in a time and place where Buddhist teachings are available, combined with a human mind and body that enable one to understand and practice the teachings, is believed to occur only rarely and is thus exceedingly valuable. Tibetan Buddhist teachings therefore emphasize the importance of cultivating a continual awareness of death in order to inspire one to make the most of one's human opportunities for spiritual progress. An oft-cited three-part Buddhist formula has one reflect on the facts that: one will definitely die; the time of death is uncertain; only spiritual training will be of any use at the time of death.

9.3 The Death Process

9.3.1 Introduction

According to Tibetan Buddhists, the moment that consciousness departs from its bodily "house" is very rarely followed immediately by rebirth. Instead, when consciousness separates from the body at death, it typically enters an intermediate state, or *bardo* (*bar do*), where it resides for as many as 49 days. The experiences of consciousness while in the *bardo* are described in a series of Tibetan texts attributed to the eighth century Guru Rinpoche (Skt. Padmasambhava) who hid them for a future time. Titled *Peaceful and Wrathful Deities: Self-Liberation through Realization* (*Zhi khro dgongs pa rang grol*), the corpus was revealed in the fourteenth century by the

Treasure Revealer (*gter ston*) Karma Lingpa (Kar ma gling pa).[4] Karma Lingpa's writings also include instructions for recognizing signs that death is approaching, directions for performing the ritual of consciousness transference (*'pho ba*), a script for a masked dance-drama (*'cham*), and short meditations for daily practice.

Within the *Peaceful and Wrathful Deities*, the text that deals with consciousness's death journey is called the *Great Liberation through Hearing in the Intermediate States* (*Bar do thos grol chen mo*), henceforth *Liberation through Hearing*. The title reflects the belief that a dying person can attain liberation by hearing the instructions read aloud at the time of death, even if the person has not engaged in spiritual practices but instead accumulated much negative karma. The *bardo* experiences described in the *Liberation through Hearing* are part of the basic religious knowledge shared by members of Tibetan Buddhist communities today. When I asked the typical age at which a child is able to describe the landscape of death and the *bardo* journey, I was told that children can do so as soon as they are able to talk! Due to the popularity of Western-language translations of the *Liberation through Hearing*, known in English as the *Tibetan Book of the Dead*, the basics of the death process are familiar to many non-Tibetans as well.

Numerous sets of instructions and interpretative commentaries about the death *bardo*s have been written since the fourteenth century. As a result, details vary according to the Tibetan lineage that one examines. Presented below is an overview based primarily on *The Mirror of Mindfulness*, a text of Dzogchen-Mahāmudrā teachings by the seventeenth century master, Tsele Natsok Rangdröl (rTse le sna tshogs rang grol).[5]

9.3.2 *The Bardos*

Tibetan tradition designates three *bardo*s that we traverse as we transition between our current human life and our rebirth into a new body: the *bardo* of dying, the *bardo* of reality, and the *bardo* of becoming. In colloquial speech, '*bardo*' refers to the entire interval between this life and the next, but in the Tibetan philosophical system, the time between the moment we are stricken with the illness that will eventually cause our death until the moment we are reborn into our next life is divided into three stages.

[4] For a history of this literature, see Cuevas (2003).

[5] The Tibetan title of the text is *Bar do spyi'i don thams cad rnam pa gsal bar byed pa dran pa'i me long*. Numerous books, both scholarly and non-scholarly, present detailed descriptions of the death experience as described in this literature. See, for example, Cuevas (2003), Sogyal Rinpoche (1993), Thondup (2005), and Thurman (1994). Coleman and Jinpa (2005) provides a complete English translation.

9.3.2.1 The *Bardo* of Dying

At the conclusion of our present life (itself a *bardo*) we arrive at the painful *bardo* of dying. Explanations of the physical changes that occur as the time of death approaches are found in Tibetan medical Tantras studied by doctors of Tibetan medicine. According to a monk doctor I interviewed, the process of dying begins with the tongue rolling back into the mouth. Then, marking the cessation of kidney function, the ears do not spring back into position when pressed against the head. Third, the nostrils expand as the lungs begin to fail. As liver function disintegrates, the eyes roll back into the head, and as the sensation of touch diminishes, the bottom lip rolls downward. Finally, when the heart stops beating, the body loses its heat and breathing ceases completely. From a Tibetan exoteric medical point of view, this marks the time of death.

As doctors have been rare and transportation slow and difficult, especially in rural areas of Tibet, senior men in the community have traditionally borne the responsibility of determining if someone has died. Some of the non-medical tests performed by laymen include evaluating the armpits and soles of the feet for lingering body heat, pinching the skin to see if it springs back, and tugging on the hair to see if it pulls out easily. Villagers might sprinkle a dusting of barley flour on the upper lip to check for shallow breathing: undisturbed flour indicates that there is no air going in or out of the nostrils. There is also a pressure point on the forearm that, when stimulated, is said to revive a person who is not truly dead.

Moving from a medical to a tantric perspective, the *Liberation through Hearing* describes a series of “dissolutions” that occur as the sense organs cease operating and consciousness coalesces at the heart-center. The speed with which the dissolution stages occur depends on the condition of the physical body and the particular manner of death. The dissolutions may last for days, months, or years, or they may be completed in the span of a few seconds.

Typically, the first dissolution is of the earth element, supporting flesh and bone, into water. This dissolution is marked by loss of physical strength. Next, the water element, comprising blood and other bodily fluids, dissolves into fire. As this happens, the mouth becomes dry, one feels nervous and irritated, and perception becomes hazy, as if one were engulfed in smoke. The third element to dissolve is usually that of fire, responsible for body heat, into wind. During this dissolution, the body loses warmth, and it is said that a person has the experience of seeing tiny lights like fireflies. Next, the wind element, comprising breath, dissolves into consciousness. Breathing becomes difficult, and the eyes roll upwards. People with good karma see welcoming visions during this stage, while those with negative karma are frightened by an apparition of the fearsome Lord of Death (gShin rje Chos rgyal). When this dissolution is complete and all sense organs have stopped functioning, the dying person is unable to see, hear, smell, taste, or feel touch. Then breathing ceases, the skin becomes pale, and only a slight bit of body heat remains at the heart. The tiny bit of warmth at the center of the chest corresponds to the movement of consciousness to the heart-center. It is important to note that although the heart has stopped beating and the lungs are no longer functioning at this point, in the Tibetan system, this is only the brink of death.

According to Thurman, a person might be considered clinically dead at this point, since "there is no movement in brain or circulatory system" (1994, p. 42), but from the Tibetan point of view, there are several more stages between clinical death and complete death.[6] It is possible, although rare, for people to remain in this liminal stage of clinical but not complete death for several hours or even days. Highly realized meditators may stay in this state for weeks in a meditation called *tukdam* (*thugs dam*).[7] Absorption in *tukdam* is an extremely noteworthy and important accomplishment and serves as final evidence of a practitioner's spiritual achievement. *Tukdam* precedes the final dissolution, the dissolution of consciousness into space, in which the dying person is freed from all mental states arising from ignorance and afforded an encounter with their innate wisdom. This natural, inborn wisdom is referred to in Tibetan as the luminosity or "clear light" of death (*bar do dang po'i 'od gsal*).[8] Tantric teachings state that a practitioner who has become familiar, through meditation, with this experience of radiant emptiness will be able to recognize it during the dissolution process and thereby achieve liberation at the moment of death.

Practices of Highest Yoga Tantra (*bla na med pa'i rgyud*, Skt. *anuttarayogatantra*) include meditations in which a yogi is able to rehearse the stages of dying. The Dalai Lama includes this type of death meditation as part of his daily practice. He claims to feel excited when he contemplates his chance, at death, to test the results of his training (*The Tibetan Book of the Dead* 1994).[9] For ordinary persons, however, the encounter with their mind's pure luminosity is overwhelming and causes them to fall into a state of unconsciousness. According to the Tibetan Buddhist tradition, only upon the cessation of the clear light of death are the dissolutions complete and consciousness fully separated from the body. As an indication that death has occurred and consciousness departed its physical form, Tibetan Buddhists look for a bit of blood and/or mucus to emerge from the nose and/or genitals. This serves as the tantric indication that a person is irrevocably and unalterably dead.

---

[6] The dissolutions are presented in abbreviated form in section 8, chapter 12 of Atiśa's *The Jewel Garland of Dialogues* (Thubten Jinpa 2008, pp. 242–51).

[7] Ling Rinpoche, one of the tutors to the Dalai Lama, remained in *tukdam* for 13 days: "Although he was clinically dead and had stopped breathing, he stayed in the meditation posture and his body showed no sign of decomposition. Another realized meditator remained in this state for 17 days in the tropical heat of high summer in eastern India" (His Holiness the Dalai Lama 2005, p. 157). In the case of the 1990 death of Khenchen Sangay Tenzin, the period of *tukdam* stretched for 9 days (Zivkovic 2014). *Tukdam* can be compared to *samādhi* at death in Hindu traditions.

[8] The appearance of this clear light is sometimes classified within the *bardo* of dying and sometimes in the beginning of the reality *bardo*. See, for example, Dzogchen Ponlop (2006, p. 140) and Thondup (2005, p. 70) where, following the Dzogchen tradition, they posit it as the first stage of the reality *bardo*.

[9] In *Kindness, Clarity, and Insight*, the Dalai Lama states that in his daily practice, he passes through the stages of dying six or seven times (2006, p. 205). Lopez 1997 translates a relevant prayer commentary composed by the Gelugpa scholar Jangkya Rolpé Dorje (Lcang skya rol pa'i rdo rje). See also Cozort (1986).

9.3.2.2 The Reality *Bardo*

The dissolution of all elements at the completion of the *bardo* of dying marks the beginning of the reality *bardo*. Most people, having fallen unconscious during the dawning of the clear light of death, awaken confusedly into the reality *bardo*. It is here that the 42 peaceful and 48 wrathful deities appear. These deities, including fearsome *heruka*s and other animal-headed deities, are much depicted in Tibetan religious art.[10] Extremely bright lights of various colors and tremendously loud sounds accompany the manifestation of the hundred deities. In actuality, all experiences of the reality *bardo* are visions produced by consciousness itself. As a result, the reality *bardo* provides advanced practitioners with further opportunities to attain liberation by recognizing the forms, colors, and sounds for what they are: projections of the mind. For most beings, however, the experience of the reality *bardo* is confusing and terrifying. They assume the intense visual and auditory displays of the reality *bardo* to be the arrival of externally existing phenomena and turn away in fear. This leads to the next *bardo*, the *bardo* of becoming.

9.3.2.3 The *Bardo* of Becoming

The *bardo* of becoming is compared to a dream. Its stages unfold in a sequence that mirrors the dissolutions of the *bardo* of dying: just as each element dissolved one into another, in the *bardo* of becoming, the elements rematerialize. This leads to the return of ignorance, desire, and anger as consciousness assumes a "mental body" (*yid lus*). The mental body possesses all sense faculties, is nourished by smells, and can travel immediately and unimpeded through space. Once again able to see and hear yet not realizing that one has died, one feels intense longing for one's possessions and angry when family and friends ignore one's requests for food and drink. In the words of a Hyolmo man I interviewed in Nepal,

> When we die, we can see everyone and what they're doing, but they don't see us. We speak to them, but they act like they don't hear us. We get angry when someone sits in our place, and we tell them that that is *our* place to sit, but they ignore us, and there's nothing we can do. We're unable to animate our [physical] body. We don't realize that we've died.

Because the mental body is immaterial, it is blown about here and there like a feather in the wind, moving wherever thoughts take it. This is described as a time of great fear and uncertainty.

In the *bardo* of becoming, the mind is believed to be many times more lucid than it was during life. One is able to read minds, and one enjoys a power of hearing that allows one to hear sounds at any distance and understand all languages. This is why, according to Sogyal Rinpoche, the oral recitation of the *Liberation through Hearing*

[10] For reproductions of paintings depicting the peaceful and wrathful deities of the reality *bardo*, see Coleman and Jinpa (2005) and especially Evans-Wentz (2008).

can be helpful for any being progressing through the *bardo* of becoming. Regardless of the language in which it is read, "the essential *meaning* of the text can be understood fully and directly by [the deceased person's] mind" (Sogyal Rinpoche 1993, p. 305, emphasis in original). Hearing the instructions as they are read aloud, one is able to comprehend that one has died and, hopefully, apply the teachings in order to take a favorable rebirth.[11]

For people who are unable to benefit from oral recitation of the teachings due, for example, to dying suddenly, there is another technique for avoiding rebirth in the lower realms. Through a ritual of consciousness transference, an accomplished meditator can direct the consciousness of a dead or dying person (or animal) away from negative rebirths towards birth in a Pure Land. Consciousness transference is a forceful method by which an adept takes charge of a being's confused wandering through the *bardo*s to send them to the land of a particular buddha or bodhisattva. The practice of consciousness transference can be undertaken for an individual at any stage of the dying process. In many cases it is repeated in tandem with readings from the *Liberation through Hearing* and merit-making rites as a sort of insurance policy for the deceased.

For most, the *bardo* of becoming lasts from 7 to 49 days. Near the midpoint of this time, one encounters a vision of the Lord of Death and his attendants. With the help of the "mirror of karma" and a scale for weighing positive and negative deeds, the Lord of Death presides over the determination of one's next birth. Finally, when the moment of rebirth arrives, beings to be conceived in a womb observe a male and female engaged in intercourse. They feel attraction towards one future parent and anger towards the other. This emotional engagement is the last moment of the *bardo* of becoming as one's karma propels one into one's future mother's womb.[12]

9.3.3 *Tibetan Buddhist Funerary Customs*

The preceding discussion explicates Chökyi Nyima Rinpoche's meaning when he writes that death is a state of "disembodied consciousness" (1991, p. 150). The potentially long duration of a dying person's journey to rebirth is reflected in Tibetan Buddhist death rituals, distinguishing them from non-Buddhist as well as Theravāda Buddhist customs.

It is standard practice among Tibetan Buddhists to retain a corpse, undisturbed, for several days before disposing of it. Bardor Tulku Rinpoche (2004) attests to a tradition in Tibet of maintaining a body for 49 days. In locations with greater heat

[11] For a look at how the instructions contained in the *Liberation through Hearing* are utilized in contemporary contexts, see "Part One: A Way of Life" in the excellent two-part documentary, *The Tibetan Book of the Dead* (1994).

[12] Beings not born from wombs take rebirth at the moment that they see their future birthplace and feel either attachment or anger.

and humidity, including contemporary exile communities in India and Nepal, this is rarely possible and 3 days is deemed sufficient. A Sherpa lama from Solu living in Kathmandu, Nepal, remembered that when Sherpas first started moving from the mountains to live in the city, they rented rooms in Hindu families' houses. If the Hindu landlords had known that one of their Sherpa tenants was deathly ill, they would have insisted that the person be carried outside to die in order to avoid ritual pollution of the house. For Tibetan Buddhists, however, touching any part of a dying person's body is hazardous because doing so is believed to draw the consciousness to that part of the body. It is likely that the consciousness will then depart from the nearest bodily orifice instead of from the crown of the head, thus compromising the person's future birth. For this reason, Sherpas kept quiet when someone was dying and performed the funeral rituals without using the traditional bells and drums. In this way, they were able to observe the important practice of leaving the body undisturbed for a minimum of 3 days.

To cremate or bury a body within a shorter period of time risks upsetting a consciousness that has not yet departed. As mentioned earlier, consciousness often does not immediately realize its change of status at the moment of death. If the body is destroyed before consciousness has understood that it is no longer connected to the body, it is believed that the consciousness will experience negative emotions, specifically fear and anger, which will negatively affect its rebirth. Bardor Tulku Rinpoche writes:

> Normally when someone dies … although they are dead and their mind is no longer biologically seated in their body the way it was when they were alive, their mind may stay in their body for a few days…. Even if it does not, they may not realize they are dead for a while, and they may therefore identify very much with the body. If the body is disturbed … they can get quite angry. And the anger that they generate at seeing what is done to their body can negatively impact the process of rebirth for them. (2004, pp. 202–203)

One Tibetan lama with whom I spoke compared destruction of the body while the consciousness remained associated with it to murder. Tsele Natsok Rangdröl echoes this view. He writes that cremation of a body before the consciousness has exited creates an obstacle for someone who is still in meditation and "an ordinary person attached to his body [whose consciousness has not moved on], will be burned alive" (1987, p. 54).

A practical reason to retain a body for a short time after death is to ensure that a corpse is indeed a corpse. Keown provides an example from the *Mahāparinibbāna-sutta* in which the Buddha entered such a deep meditative trance that he stopped breathing at which point his disciple Ānanda was unable to determine whether he was alive or dead (2010, p. 10). Cessation of respiration is understood to occur even at lower levels of meditative absorption, such as in the fourth *jhāna* (Keown 2010, p. 21). In the absence of sophisticated medical equipment, it is therefore possible to mistakenly declare someone dead.

9.3.4 Returning from Death

Retaining a corpse for several days after death also allows for the extremely rare possibility that a person, having died, will return to life. Chökyi Nyima Rinpoche writes, "There have been rare cases where the consciousness, having left the body, actually returns to the corpse, revives it and continues to live. In Tibetan this is called returning from the dead" (1991, p. 153). People who return from death are known in Tibetan as *delog*s (*'das log*), literally "passed away and returned."[13] The *delog* phenomenon is different from a near-death experience in that a *delog* not only comes near to death but dies completely. In Bardor Tulku Rinpoche's words, *delog*s "actually go all the way… really die, and after up to a week or more come back to their bodies" (2004, p. 203). Another way that a *delog* experience differs from a near-death experience is the length of time that a *delog* remains dead. Describing his mother, Dawa Drolma's, *delog* journey, Chagdud Tulku takes care to make this clear: "Hers was not a visionary or momentary near-death experience. For five full days she lay cold, breathless, and devoid of any vital signs, while her consciousness moved freely into other realms" (Dawa Drolma 1995, p. vii).

Accounts of individuals who have experienced and then returned from death support Tibetan Buddhist understandings of the death process. Autobiographies of *delog*s and interviews I conducted with *delog*s in Tibet and Nepal assert that the *delog*'s consciousness exited the body which then became indistinguishable from a corpse. A *delog* in the rural Tibetan region of Golok, for instance, had her body examined by several doctors who confirmed the complete cessation of breath. "A Chinese doctor from the Riwong clinic listened with a stethoscope," the *delog* told me. "A famous Tibetan doctor felt my pulse. They both said that I wasn't breathing." She continued to describe how the blood stopped flowing through her veins and her awareness left her body.

During their death adventures, *delog*s witness beings in the *bardo* of becoming having their positive and negative karma evaluated in front of the Lord of Death. Then, instead of progressing to their next life, *delog*s are instructed to return to the human realm. They do so with good and bad news about the dead whom they encountered as well as messages from the deceased and the Lord of Death. The re-entrance of a *delog*'s consciousness into the body is described as painful. A *delog* in Amdo told me of the fear she felt when she revived to find that her children had changed her bedding, making it difficult for her to locate her body. For this reason, butterlamps are often burned as a guide for the consciousness if it returns to reinhabit the physical form. They are placed near the head of the recently deceased, as consciousness is said to return via the location of the anterior fontanelle.

[13] For a book-length treatment of the *delog* phenomenon based on textual sources as well as interviews with *delog*s in Bhutan, see Pommaret (1989). A brief account of a Tibetan *delog* can be found in Prude (2014).

The *delog* phenomenon raises interesting questions about the possibility of returning from death. Some skeptics insist that a *delog* does not completely die. In their view, the *delog* journey is a dream or a hallucination caused by illness or some other bodily stress. The Dalai Lama suggests that *delog*s, as well as people who undergo near-death experiences, do not enter the death *bardo*s but a different *bardo*, the *bardo* of dreaming. He believes that if a consciousness has fully separated from the body in the way that it does at death, it would have no memory of its experience in the *bardo*s were it to return to the body. Based on the *delog*'s ability to remember her *bardo* experiences, he concludes that the *delog*'s consciousness must not completely leave her body (The Dalai Lama 1997). Other Tibetan Buddhist scholars disagree. According to Lati Rinpoche, if the cause of a person's death is "some accidental circumstance, not a karmic debt," then it is possible for a person to be revived from death "by medical or spiritual means" (Hopkins and Lati Rinpoche 1979, p. 49). This latter viewpoint relies on the idea of timely versus untimely death to explain seemingly miraculous returns to life.

9.3.5 Timely Versus Untimely Deaths

Tibetan writings about timely and untimely death are included in Karma Lingpa's *Peaceful and Wrathful Deities*, and the distinction is familiar to Tibetan Buddhists.[14] A timely death is one that occurs in accordance with the exhaustion of one's lifespan (*tshe*) as determined by past karma and the maintenance of life-sustaining merit (*bsod nams*). Timely deaths cannot be postponed or avoided. Deaths due solely to the depletion of merit, however, may be prevented. A professor in the Amdo region of Tibet explained: "Throwing out food that we don't want to eat is a root cause of exhausting our merit. If we always need to wear new clothes and not our old ones, this exhausts our merit. If we must always live in a nice house and not in an old or run down one, we will exhaust our merit." Other causes he listed include not respecting our parents and elders, not taking care of the young, and not being honest and friendly towards our friends. Behaviors disrespectful of Buddhist monastics, such as sitting on a monk's robe, also diminish one's merit. To rebuild one's store of merit, Tibetan Buddhists perform rituals such as prostration, pilgrimmage, saving the lives of animals (*tshe thar*), *maṇḍala* offering, and *mantra* recitation. They may also sponsor monks to recite supporting prayers on their behalf.

In order to live out one's natural lifespan, a person must also care for their health and avoid dangerous situations. Attending long-life rituals that may involve offering sacrificial cakes (*gtor ma*) and imbibing consecrated "nectar" (*bdud rtsi*) and blessing pills (*ril bu*) provide further protection from obstacles to the lifespan. The deities Amitāyus, White Tārā, and Uṣṇīṣavijayā (rNam rgyal ma) are known as the

[14] These ideas can also be found in Indian Buddhist Abhidharma literature, such as the Buddhist philosopher Asaṅga's (c. fourth century) *Abhidharmasamuccaya*. Mullin (1998) discusses untimely death from a Tibetan Buddhist perspective.

three deities of long life and prayed to for this purpose. Tibetan lore is full of accounts of devotees who were able to extend their lives by, for example, commissioning statues and paintings of particular deities.[15] In addition to the prophylactic behaviors just mentioned, a yogi who has performed the necessary preliminaries can learn to recognize outer, inner, and secret signs that portend death.[16] When an indication, such as a dream of being surrounded by crows, signals that death is approaching, the longevity rituals mentioned above may be effective at adverting an untimely death. Untimely deaths caused by obstacles such as demons or black magic may be avoided by engaging in rituals that "cheat death" (*'chi bslu*). Some of these rituals are ransom rites that offer a substitute, such as an effigy of the human body, to the negative forces that seek to steal the life-force (*srog*) and thus cause one to die before the completion of one's lifespan.[17]

In case an untimely death is not avoided, there exists a possibility for it to be reversed. According to a lama from the Chamdo region of Tibet, if "the soul (*bla*) has not deteriorated, the lifespan is not exhausted, but the life-force (*srog*) has been severed," when the consciousness encounters the Lord of Death in the *bardo* of becoming, it will be sent back to the human realm to complete the tenure of its life. In the biography of sixteenth century *delog* Lingsa Chökyi, we read: "I went [into his presence], and the Dharma King [Lord of Death] said to me, 'Ah, your lifespan is not finished. Your name and identity were mistaken, and you were brought here. Because your aggregates [i.e., body] are still in your bed, return!'" (Ngag dbang blo bzang bstan 'dzin rgya mtsho 2001, p. 78).[18] From a Buddhist perspective, a human mind and body are the ideal supports for achieving liberation, and one cannot know when or where one will attain another precious human birth. It follows that a person should employ all available means to avoid premature death.

9.3.6 Dealing with Death in the Context of Modern Medicine

Doctors of Tibetan medicine are trained in tantric Buddhist theories as well as anatomy and pharmacology, so traditionally there was little disconnect between the physical and spiritual care that Tibetan Buddhists received. Recommended medical treatment could take the form of rituals performed on the patient's behalf by

[15] The recipient of numerous and repeated long-life prayers and offerings, the Fourteenth Dalai Lama has reported that he expects to live to be at least 100 years old, and based on his dreams and "other indications," he may live to the age of 113 (Central Tibetan Administration 2017). On a long-life initiation focused on the bodhisattva Tārā, see Beyer (1978), pp. 375–398.

[16] Germano (1997) includes a list of dream signs indicating impending death. See also the relevant chapter in *The Tibetan Book of the Dead* (Coleman and Jinpa 2005).

[17] Walter (2000) translates sections of the *Teaching on Cheating Death* (*'Chi ba bslu ba'i man ngag*) an originally Sanskrit text incorporated into the *Tengyur*. For descriptions, classifications, and a history of these rituals, see Mengele (2010).

[18] Cuevas (2008, p. 52) translates this section of text based on a different version of Lingsa Chökyi's *delog* narrative.

Buddhist monks and meditators. This is still the case today. In addition, many people living in the Himalayas and on the Tibetan Plateau lack access, whether for geographic or economic reasons, to state-of-the-art hospitals and medical equipment. But for Tibetan Buddhists who are able to make use of modern medical procedures, how do religious ideas about death and dying impact end-of-life decisions?

Answers to contemporary bioethical dilemmas cannot be found in canonical texts; they must be improvised. Furthermore, within Tibetan Buddhist traditions, there is no single central authority that dictates regulations for all to follow. In fact, in an appendix to *The Tibetan Book of Living and Dying*, written for a western audience, Sogyal Rinpoche warns against taking an "'official' position" or "mak[ing] rules" to govern end-of-life dilemmas (1993, p. 371). Instead, in accordance with the opinion of the Dalai Lama, he suggests attending to the particulars of each situation while acting with as much compassion as possible. In Tibet and the Himalayas, a variety of doctors, Buddhist teachers, and ritualists may be consulted about difficult medical decisions, and divination techniques may be used to determine the best course of action.

When prognostications indicate certain death, Tibetan Buddhists might not pursue medical treatment at all, especially if the treatment is costly.[19] The *Kālachakra Tantra* warns against intervening when signs indicate that death is unavoidable because attempting to heal a patient who is sure to die is useless, and treating dying patients could tempt medical professionals to put their own monetary interests before the wellbeing of their patients (Wallace 1995). It is therefore extremely important to be able to distinguish between cases in which life can be meaningfully extended and cases in which it cannot. Instead of attempting to avoid death for as long as possible, Buddhist teachings would have people accept the certainty of death and learn to face death with less anxiety. Verses in Atiśa's (982–1054) *The Jewel Garland of Dialogues* impart a theme common to Tibetan Buddhist teachings: mindfulness of death.

> Some die in old age;
> Some die when young
> Some die in the prime of youth;
> Some die just after birth.
> Male or female, intelligent or learned,
> Who has seen or heard of anyone who was spared? (Thubten Jinpa, p. 208)

Reflecting on the inevitability of death is intended to inspire a higher quality of life, a factor Buddhist teachings posit as more valuable than a long lifespan.[20]

Aggressive treatments at the end of life, especially, are seen as pointless in addition to spiritually dangerous as they can increase negative attitudes like attachment

---

[19] See, for example, Schrempf (2011).

[20] In a book written for medical professionals, Chokyi Nyima Rinpoche indentifies being able to alleviate the fear of illness and dying as "the greatest help… the nicest way to help someone else" (2008, p. 31). For their part, family members are encouraged to remember Buddhist teachings and not pursue medical care for their loved one at the expense of the loved one's spiritual well-being.

and frustration in the person who is dying. Tibetan Buddhist doctrines identify one's state of mind at death as an important determinant of one's rebirth. Thoughts of anger at the time of death can lead to rebirth in hell, for example, while meditation on a buddha's Pure Land can result in rebirth there. For these reasons, a rush of hospital staff and noisy equipment are seen as problematic due to their potential to inhibit a calm and peaceful transition between lives. The same can be true of emergency life-saving techniques, such as ventillators and CPR. Tsomo writes, "From a Tibetan Buddhist perspective, to pound on a dying person's chest in an attempt to resuscitate the heart function at the time of death is probably the worst thing that can be done" because for patients who do not revive, the aggressive nature of CPR may be frightening and upsetting (Tsomo 2006, p. 218). Instead of employing aggressive life-saving measures, when death is determined to be inevitable, Tibetan techniques emphasize palliative care. The goal is to make the dying person as comfortable as possible so that they can make best use of the conclusion of their human birth.

The Tibetan tradition accords with general Buddhist precepts that prohibit active euthanasia.[21] Hastening death in the context of terminating physical pain is seen as ultimately ineffective because the karma responsible for the painful experience will only carry over into a future life. The desire to end a human life, whether one's own or someone else's, is viewed negatively because it is based in ignorance of the fact that death does not bring an end to the negative karma that causes suffering.[22] In addition, pain, however unpleasant, can serve a soteriological purpose by reminding one of the First Noble Truth, the truth of *duḥkha*, and the basic fact of impermanence. A premature death can thus cut short one's chances to engage the painful *bardo* of dying as part of one's spiritual path. Instead, analgesics that do not cloud the mind or impair judgment may be administered and Buddhist practitioners called to perform rituals to purify the dying person's negative karma.

This issue of voluntary, passive euthanasia, on the other hand, seems to find support among contemporary Tibetan Buddhist teachers. The Dalai Lama writes, "If the brain is not functioning, and if only the body is kept alive at great expense, it would be more useful to spend the money for some other purpose, assuming that there is utterly no hope of recovery" (The Dalai Lama 1997, p. 162). Similarly, Dilgo Khyentse Rinpoche condones ceasing life-support measures when "there is no hope" (Sogyal Rinpoche 1993, p. 372). In fact, if, through meditation, a person perceives that they will die in a few months, it is acceptable to perform the transference of consciousness ritual and exit the body early because, in the words of the Dalai Lama, "letting your body deteriorate in sickness would make it very difficult to meditate and exit from this life properly" (The Dalai Lama 1997, p. 174). Again, we see that one's quality of life and potential to cultivate positive states of mind are valued more highly than the number of hours and days that a person is able to live.

---

[21] For a discussion of the ethical problems inherent in euthanasia, see Tsomo (2006, p. 178) and Harvey (2000, pp. 292–310).

[22] Kalu Rinpoche interprets a terminal patient's stated desire to end their life as actually a desire to relieve suffering and thus karmically neither positive nor negative (Sogyal Rinpoche 1993, p. 374).

The question of organ donation is more complicated. Some Tibetan Buddhist masters have spoken in support of organ donation because the altruistic wish to give away one's unneeded corpse generates positive karma (Sogyal Rinpoche 1993, pp. 376–77). On the other hand, because organ harvesting is undertaken before the body is fully dead, some Tibetan Buddhists have chosen not to be organ donors out of concern that having their body cut into before the completion of the death process would interfere with their efforts to realize the clear light of death (Vesna Wallace, personal communication, 12/15/17). Keown cites an anonymous "authority" supporting the view that "'from a Tibetan point of view organ harvesting done within three days after the stoppage of the heart is basically the same as cutting organs out of a living being'" (2010, p. 3). This assessment stems from the inability of medical equipment which assesses brain function to confirm the absence of the subtlest level of consciousness.[23] As long as this subtle consciousness is present in the body, the dying person's future can be influenced by his or her state of mind.

9.3.7 Reflections

The dissolution of the elements of the physical body at death presents a remarkable spiritual opportunity. If one overcomes the fear generated by the unfamiliar experiences and succeeds in attaining awareness of what is happening, one will attain liberation from *saṃsāra*. From a Tibetan Buddhist perspective, a good death is thus more valuable than a prolonged life.

As an exploration of Tibetan Buddhist approaches to death and dying makes clear, death is more than a matter of medical diagnosis. Death poses a hermeneutic question. The Tibetan Buddhist idea of untimely death allows for the possibilities both of averting and returning from death, and it is often the case that consciousness departs the body slowly, in stages, thus confounding attempts to identify a precise "moment" of death. Even after all vital signs have ceased, such as during post-death *tukdam* meditation, Tibetan philosophies posit a subtle consciousness that may remain with the body. Interestingly, while a practitioner is absorbed in *tukdam*, a slight bit of warmth is detectable at the heart and the skin remains pliant. Only when *tukdam* is complete does the body begin exhibiting indicators of death, such as pallor and rigor mortis, suggesting that there may be some degree of life that continues after clinical death.

---

[23] In a compelling argument against using brain death as a criterion for death, Keown points out:

> According to Buddhist teachings, mental awareness (*mano-viññāna*) is not the essence of a human being … and is only one of six forms of consciousness diffused throughout the human body. Even though mental awareness may be lost temporarily or permanently, this does not mean that the deeper underlying forms of organic consciousness or *viññāna*—which I prefer to translate as "sentiency"—do not continue. (2010, p. 19)

Tibetan Buddhist understandings of the death process bridge scientific, materialist observations and religio-spiritual interpretations. In contrast to Western approaches that tend to pursue life at all costs, Tibetan Buddhist attitudes emphasize constant recollection of death in order to diminish the shock and resistance one experiences when death arrives. As opposed to "medicalizing" death, it could be argued that, through the use of elixirs, long-life pills, and lifepan-increasing rituals and acts of merit, Tibetan Buddhist practices "medicalize life."

References

Bardor Tulku Rinpoche. 2004. Preparing for death and the bardo states. In *Rest for the fortunate*. Trans. Lama Yeshe Gyamtso. 171–215. Kingston, NY: Rinchen Publications.

Beyer, Stephan. 1978. *The cult of Tārā: Magic and ritual in Tibet*. Berkeley: University of California Press.

Central Tibetan Administration. 2017. *I will live for more than 100 years: His Holiness reaffirms at final day of Kalachakra*. January 16. http://tibet.net/2017/01/i-will-live-for-more-than-100-years-his-holiness-reaffirmed-on-final-day-of-kalachakra/. Accessed 8 Sept 2017.

Chökyi Nyima Rinpoche. 1991. *The bardo guidebook*. Trans. Erik Pema Kunsang, ed. Marcia Schmidt. Hong Kong: Rangjung Yeshe Publications.

Chökyi Nyima Rinpoche, with David Shlim. 2008. *Medicine and compassion: A Tibetan lama's guidance for caregivers*. Trans. Erik Pema Kunsang. Boston: Wisdom Publications.

Coleman, Graham, and Thupten Jinpa, eds. 2005. *The Tibetan book of the dead*. Trans. Gyurme Dorje. New York: Penguin.

Cozort, Daniel. 1986. *Highest Yoga Tantra: An introduction to the esoteric Buddhism of Tibet*. Ithaca: Snow Lion.

Cuevas, Bryan. 2003. *The hidden history of the Tibetan book of the dead*. New York: Oxford University Press.

———. 2008. *Travels in the netherworld: Buddhist popular narratives of death and the afterlife in Tibet*. New York: Oxford University Press.

Dawa Drolma, Delog. 1995. *Delog: Journey to realms beyond death*. Trans. Richard Barron. Junction City: Padma Publishing.

Dzogchen Ponlop. 2006. *Mind beyond death*. Ithaca: Snow Lion.

Evans-Wentz, Walter, ed. 2008. *Tibetan book of the dead: An illustrated edition of the sacred text on death and rebirth*. New York: Metro Books.

Fenner, Todd. 1996. The origin of the *rGyud bzhi*: A Tibetan medical tantra. In *Tibetan literature: Studies in genre*, ed. José Cabezón and Roger Jackson, 458–469. Ithaca: Snow Lion.

Garrett, Francis. 2009. The alchemy of accomplishing medicine (*sman sgrub*): Situating the *Yuthog Heart Essence* (*G.yu thog snying thig*) ritual tradition. *Journal of Indian Philosophy* 37 (3): 207–230.

Germano, David. 1997. Dying, death, and other opportunities. In *Religions of Tibet in practice*, ed. Donald S. Lopez, Jr., 458–493. Princeton: Princeton University Press.

Gyatso, Janet. 2004. The authority of empiricism and the empiricism of authority: Medicine and Buddhism in Tibet on the eve of modernity. *Comparative studies of South Asia, Africa and the Middle East* 24 (2): 83–96.

Harvey, Peter. 2000. *An introduction to Buddhist ethics*. Cambridge, UK: Cambridge University Press.

His Holiness the Dalai Lama. 1997. *Sleeping, dreaming, and dying: An exploration of consciousness with the Dalai Lama*, ed. Francisco Varela. Boston: Wisdom Publications.

———. 2005. *The universe in a single atom: The convergence of science and spirituality*. New York: Morgan Road Books.

______. 2006. *Kindness, clarity, and insight*. Trans. and ed. by Jeffrey Hopkins. Ithaca: Snow Lion.

Hopkins, Jeffrey, and Lati Rinpoche. 1979. *Death, intermediate state and rebirth in Tibetan Buddhism*. Ithaca: Snow Lion.

Keown, Damien. 2010. Buddhism, brain death, and organ transplantation. *Journal of Buddhist Ethics* 17: 1–34.

Lopez, Donald S., Jr. 1997. A prayer for deliverance from rebirth. In *Religions of Tibet in practice*, ed. Donald S. Lopez, Jr., 442–457. Princeton: Princeton University Press.

Mengele, Irmgard. 2010. Chilu ('Chi bslu): Rituals for "deceiving death." In *Tibetan ritual*, ed. José Ignacio Cabezón, 103–129. New York: Oxford University Press.

Mullin, Glen. 1998. *Living in the face of death*. Ithaca: Snow Lion.

Ngag dbang blo bzang bstan 'dzin rgya mtsho. 2001. *'Das log gling sa chos skyid kyi rnam thar*. (*The Biography of Delog Lingsa Chökyi*.) Delhi: Chos spyod Publications. [In Tibetan.]

Pommaret, Françoise. 1989. *Les revenants de l'au-delà dans le monde tibétain*. Paris: C.N.R.S. [In French.]

Prude, Alyson. 2014. Kunzang Drolkar: A *Delog* in eastern Tibet. In *Eminent Buddhist women*, ed. Karma Lekshe Tsomo, 169–184. Albany: SUNY University Press.

Schrempf, Mona. 2011. Between mantra and syringe: Healing and health-seeking behaviour in contemporary Amdo. In *Medicine between science and religion: Explorations on Tibetan grounds*, ed. Vincanne Adams, Mona Schrempf, and Sienna Craig, 157–183. New York: Berghahn Books.

Sogyal Rinpoche. 1993. *The Tibetan book of living and dying*. San Francisco: Harper.

The Tibetan Book of the Dead. 1994. DVD. Directed by Barrie McLean. Montreal: National Film Board of Canada.

Thondup, Tulku. 2005. *Peaceful death, joyful rebirth*. Boston: Shambhala.

Thubten Jinpa, trans. 2008. The book of Kadam: The core texts. Boston: Wisdom.

Thurman, Robert, trans. 1994. *The Tibetan book of the dead*. New York: Bantam.

Tsele Natsok Rangdröl. 1987. *The mirror of mindfulness*. Trans. Erik Pema Kunsang. New Delhi: Rupa.

Tsomo, Karma Lekshe. 2006. *Into the jaws of Yama, Lord of Death: Buddhism, bioethics, and death*. Albany: SUNY Press.

Wallace, Vesna. 1995. Buddhist tantric medicine in the *Kālacakratantra*. *Pacific World* 11: 155–174.

Walter, Michael. 2000. Cheating death. In *Tantra in practice*, ed. David Gordon White, 605–623. Princeton: Princeton University Press.

Zivkovic, Tanya. 2014. *Death and reincarnation in Tibetan Buddhism*. New York: Routledge.

Part III
Bioethics and Religion

Chapter 10
Jewish Perspectives on End-of-Life Decisions

Elliot N. Dorff

Abstract This article first examines six fundamental Jewish convictions that affect end-of-life care. It then discusses Advance Directives. This is followed by an extensive section on the details of end-of-life care as from the perspective of Jewish law, tradition, and theology. This includes defining death, foregoing life-sustaining treatment, artificial nutrition and hydration, curing the patient and not the disease, pain control and palliative care, medical experimentation and research, and social support of the sick. The last section discusses care of the deceased, including Jewish norms about burial, cremation, autopsies, organ and tissue donation, and donating one's body to science.

10.1 Fundamental Jewish Beliefs Concerning Health Care

Judaism's positions on issues in health care stem from three of its underlying principles: that the body belongs to God; that human beings have both the permission and the obligation to heal; and that the physician holds authority in decisions about health care.[1]

10.1.1 God's Ownership of Our Bodies

According to Jewish sources, God owns everything, including our bodies.[2] God loans them to us for the duration of our lives, and they are returned to God when we die. The immediate implication of this principle is that neither men nor women have

[1] In Dorff (1998, pp. 14–34), I describe seven foundational principles for Jewish medical ethics, but these three will suffice for the purposes of this topic.

[2] See, for example, Deuteronomy 10:14; Psalms 24:1. See also Genesis 14:19, 22 (where the Hebrew word for "Creator" [*koneh*] also means "Possessor," and where "heaven and earth" is a

E. N. Dorff (✉)
American Jewish University, Los Angeles, CA, USA
e-mail: edorff@aju.edu

T. D. Knepper et al. (eds.), *Death and Dying*, Comparative Philosophy of Religion 2, https://doi.org/10.1007/978-3-030-19300-3_10

the right to govern their bodies as they will; God can and does assert the right to restrict the use of our bodies according to the rules articulated in Jewish law.

One set of rules requires us to take reasonable care of our bodies. That is why a Jew may not live in a city where there is no physician (J. *Kiddushin* 66d; cf. B. *Sanhedrin* 17b).[3] It is also the reason that rules of good hygiene, sleep, exercise, and diet are not just recommendations but commanded acts that we owe God. So, for example, bathing is a commandment (*mitzvah*) according to Hillel, and Maimonides includes his directives for good health in his code of law, making them just as obligatory as other positive duties like caring for the poor (Hillel: *Leviticus Rabbah* 34:3; Maimonides: M.T. *Laws of Ethics (De'ot)*, chaps. 3–5).

Just as we are commanded to take positive steps to maintain good health, so are we obligated to avoid danger and injury (B. *Shabbat* 32a; B. *Bava Kamma* 15b, 80a, 91b; M.T. *Laws of Murder* 11:4–5; S.A. *Yoreh De'ah* 116:5 gloss; S.A. *Hoshen Mishpat* 427:8–10). Indeed, Jewish law views endangering one's health as worse than violating a ritual prohibition (B. *Hullin* 10a; S.A. *Orah Hayyim* 173:2; S.A. *Yoreh De'ah* 116:5 gloss). So, for example, anyone who cannot subsist except by taking charity but refuses to do so out of pride is shedding blood and is guilty of a mortal offense (S.A. *Yoreh De'ah* 255:2). Similarly, Conservative, Reform, and some Orthodox authorities have prohibited smoking as an unacceptable risk to our God-owned bodies (Bleich 1977; Freehof 1977, chap. 11; *Proceedings* 1983, p. 182; all reprinted in Dorff and Rosett 1988, pp. 349–359).

Ultimately, human beings do not have the right to dispose of their bodies at will (that is, commit suicide), for that would be a total obliteration of that which belongs to God (Genesis 9:5; M. *Semahot* 2:2; B. *Bava Kamma* 91b).[4] In Judaism the theoretical basis for this prohibition is clear; we do not have the right to destroy what is not ours.

---

merism for those and everything in between); Exodus 20:11; Leviticus 25:23, 42, 55; Deuteronomy 4:35, 39; 32:6.

[3] In this and all of the following citations, these are the abbreviations that refer to rabbinic materials:

- M. = Mishnah, edited by Rabbi Judah Ha-Nasi, c. 200 C.E.
- J. = Jerusalem Talmud, edited c. 400 C.E.
- B. = Babylonian Talmud, edited c. 500 C.E.
- M.T. = Maimonides' *Mishneh Torah,* completed in 1177 C.E.
- S.A. = Joseph Karo's *Shulḥan Arukh,* completed in 1563, with glosses later inserted by Moses Isserles to indicate where the practices of Ashkenazic Jewry in northern and eastern Europe differed from the practices of Sephardic Jewry in the Mediterranean basin, which Karo had recorded.

[4] *Genesis Rabbah* 34:19 states that the ban against suicide includes not only cases where blood was shed, but also self-inflicted death through strangulation and the like; M.T. *Laws of Murder* 2:3; M.T. *Laws of Injury and Damage* 5:1; S.A. *Yoreh De'ah* 345:1–3. See Bleich (1981), chap. 26, cf. Dorff (1998, pp. 176–198 and pp. 375–376), where the official statement on assisted suicide of the Conservative Movement's Committee on Jewish Law and Standards is reprinted (also in Mackler 2000, pp. 405–434) and at https://www.rabbinicalassembly.org/sites/default/files/assets/public/halakhah/teshuvot/19912000/assistedsuicide.pdf.

10.1.2 The Human Duty to Try to Heal Ourselves and Others

God's ownership of our bodies is also behind our obligation to help other people escape pain, sickness, injury, and death (*Sifra* on Leviticus 19:16; B. *Sanhedrin* 73a; M.T. *Laws of Murder* 1:14; S.A. *Hoshen Mishpat* 426). It is not for some general (and vague) humanitarian reason or for reasons of anticipated reciprocity. Even the duty of physicians to heal the sick is not a function of a special oath they take, an obligation of reciprocity to the society that trained them, or a contractual promise that they make in return for remuneration. It is because all creatures of God are under the divine imperative to help God preserve and protect what is His.

This is neither the only possible conclusion, nor the obvious one, from the Bible. Since God announces Himself as our healer in many places in the Bible (e.g., Exodus 15:26; Deuteronomy 32:39; Isaiah 19:22; 57:18–19; Jeremiah 30:17; 33:6; Hosea 6:1; Psalms 103:2–3; 107:20; Job 5:18), perhaps medicine is an improper human intervention in God's decision to inflict illness or bring healing, indeed, an act of human hubris.

The classical Rabbis of the Talmud and Midrash were aware of this line of reasoning, but they counteracted it by pointing out that it is God who authorizes us and, in fact, requires us to heal. They found that authorization and imperative in two biblical verses. According to Exodus 21:19–20, an assailant must insure that his victim is "thoroughly healed," and Deuteronomy 22:2 requires the finder to "restore the lost property to him." The Talmud understands the Exodus verse as giving *permission* for the physician to cure (B. *Bava Kamma* 85a). On the basis of an extra letter in the Hebrew text of the Deuteronomy passage, the Talmud declares that that verse includes the *obligation* to restore another person's body as well as his or her property, and hence there is an obligation to come to the aid of someone else in a life-threatening situation (B. *Bava Kamma* 81b; see also *Sifrei Deuteronomy* on Deuteronomy 22:2 and *Leviticus Rabbah* 34:3). On the basis of Leviticus 19:16 ("Nor shall you stand idly by the blood of your fellow"), the Talmud expands the obligation to provide medical aid to encompass expenditure of financial resources for this purpose (B. *Sanhedrin* 73a, 84b, with Rashi's commentary there). Rabbi Moses ben Naḥman (Naḥmanides, 1194–1270) understands the obligation to care for others through medicine as one of many applications of the Torah's principle, "And you shall love your neighbor as yourself" (Leviticus 19:18).[5]

Medical experts, in turn, have special obligations because of their expertise. Thus Rabbi Joseph Karo (1488–1575), the author of one of the most important Jewish codes, says this:

> The Torah gave permission to the physician to heal; moreover, this is a religious precept and is included in the category of saving life, and if the physician withholds his services, it is considered as if he is shedding blood. (S.A. *Yoreh De'ah* 336:1)

[5] Naḥmanides (1963, p. 43); this passage comes from Naḥmanides' *Torat Ha'adam (The Instruction of Man), Sh'ar Sakkanah (Section on Danger)* on B. *Bava Kamma*, chapter 8, and is cited by Joseph Karo in his commentary to the *Tur, Bet Yosef, Yoreh De'ah* 336. Naḥmanides bases himself on similar reasoning in B. *Sanhedrin* 84b.

The following rabbinic story indicates that the rabbis recognized the theological issue involved in humans engaging in medical care, but it also expresses the clear assertion of the Jewish tradition that the physician's work is legitimate and, in fact, obligatory:

> Just as if one does not weed, fertilize, and plow, the trees will not produce fruit, and if fruit is produced but is not watered or fertilized, it will not live but die, so with regard to the body. Drugs and medicaments are the fertilizer, and the physician is the tiller of the soil. (*Midrash Temurrah* as cited in Eisenstein 1915, pp. 580–581)[6]

This is a remarkable concept, for it declares that God does not bring about all healing or creativity on His own, but rather depends upon us to aid in the process and commands us to try to help Him. We are, in the Talmudic phrases, God's agents and partners in the ongoing act of creation (B. *Shabbat* 10a, 119b.).[7]

At the same time, we must recognize that our life span is limited. This is the brunt of the Garden of Eden story, in which Adam and Eve eat of the fruit of the Tree of Knowledge of good and evil but are banished from the Garden before they can eat of the Tree of Life (Genesis 3). It is also the point that Kohelet (Ecclesiastes) makes when he says (3:2), "There is a time to be born and a time to die." Medical care at the end of life, then, involves the tricky decision of when physicians should do what they can to extend life and when they should let nature take its course.

10.1.3 The Relative Authority of the Physician and Patient

Because the body belongs to God, each person is duty-bound to seek both preventive and curative medical care and to follow the expert's advice in preserving one's health. Physicians, in turn, are required to elicit the patient's cooperation by making sure that the patient understands and agrees to the therapy. When several forms of therapy are medically legitimate but offer different benefits and burdens, the patient has the right to choose which regimen to follow, as long as it fits within the rubric of Jewish law (B. *Bava Metzia* 85b).[8]

---

[6] Cf. B. *Avodah Zarah* 40b, a story in which Rabbi expresses appreciation for foods that can cure. Although circumcision is not justified in the Jewish tradition in medical terms, it is instructive that the Rabbis maintained that Jewish boys were not born circumcised specifically because God created the world such that it would need human fixing, a similar idea to the one articulated here on behalf of physicians' activity despite God's rule; see *Genesis Rabbah* 11:6; *Pesikta Rabbati* 22:4.

[7] In the first of those passages, it is the judge who judges justly who is called God's partner; in the second, it is anyone who recites Genesis 2:1–3 (about God resting on the seventh day) on Friday night who thereby participates in God's ongoing act of creation. The Talmud in B. *Sanhedrin* 38a specifically objected to the Sadducees saying that angels or any being other than humans participate with God in creation.

[8] On this subject generally, see Reisner (1991a, pp. 60–62) (in Mackler 2000, pp. 250–253) and at https://www.rabbinicalassembly.org/sites/default/files/assets/public/halakhah/teshuvot/19861990/reisner_care.pdf, pp. 10–13.

On the other hand, patients do not have the right to demand of their physicians forms of treatment that, in the judgment of the physicians, are medically unnecessary, unwise, or futile or that violate their own understanding of Jewish law. That is, physicians are just as much full partners in medical care as are patients. So, for example, if a dying patient asks for interventions that physicians think are futile, his or her doctor need not, and probably should not, comply with the patient's wishes.

10.1.4 Institutional Authority and Individual Conscience

The Jewish tradition, perhaps more than any other, has used legal methods to make moral decisions. The underlying Jewish belief is that God declared His will at Sinai and specifically commanded that we not add to it nor detract from it legislatively but that we do apply it to concrete situations judicially (Deuteronomy 4:2; 13:1; 17:8–13; see also Exodus 18 and Deuteronomy 1:9–18). The rabbinic tradition understood that judicial mandate broadly, with the result that rabbinic law is much more voluminous and detailed than biblical law is. The Torah (the Five Books of Moses), in other words, is the constitution of the Jewish people, and rabbinic interpretations and rulings function as legislation and judicial rulings do in American law. Custom is also an important source of Jewish law.[9]

Most decisions that Americans would call "moral," then, are part and parcel of the legal system in Judaism. So, for example, if one wanted to know whether it is moral to withdraw life-support systems, one would ask one's rabbi, the local expert in Jewish law, and he (or she, in recent decades) would look up the question in the legal resources of the Jewish tradition. If there is some disagreement among previous or contemporary rabbis who have ruled on such cases, or if there are complications in the specific case at hand, the rabbi would use standard legal methods in deciding that specific case. The rabbi might also consult another rabbi with acknowledged expertise in the area. The lay Jew, then, would follow the ruling of his or her rabbi for both communal and theological reasons—that is, because the person wants to live within the rules of his or her community and because ultimately the Jewish tradition sees the entirety of Jewish law as commanded by God.

This methodology still holds for Orthodox and Conservative Jews, at least in theory and often in practice, for both of those branches of Judaism hold that Jewish law is binding. The Reform movement, however, champions individual autonomy, and so moral decisions are totally a matter of what the individual thinks is right. He or she may consult a rabbi, but the rabbi's words will not be authoritative law but an individual's advice—albeit an individual with expertise in the Jewish tradition.

The Talmud asserts that there are also moral norms beyond the limits of the law (B. *Berakhot* 7a; B. *Ketubbot* 97a; B. *Bava Metzi'a* 24b, 30b; see M.T. *Foundations of the Torah* 5:11; M.T. *Human Dispositions* 1:5) that we are to obey, just as God

[9] See Dorff and Rosett (1988) for more on the sources, methods, and guiding beliefs of Jewish law.

does (B. *Avodah Zarah* 4b). Such moral norms are as binding as the law is—and, indeed, the Talmud says that the Second Temple was destroyed because the Jews at the time obeyed only the laws and not the moral norms beyond the letter of the law (B. *Bava Metzi'a* 30b). As a result, even those who conscientiously abide by Jewish law might feel moral imperatives beyond what the law requires. For that matter, the rabbi might rule on the basis of such imperatives in addition to the specific sources of the law, for ultimately we are commanded to "do what is right and good in the eyes of the Lord" (Deuteronomy 6:18).[10] Jewish moral norms are defined largely by Jewish law, but also by Jewish stories, proverbs, history, family and community life, and theology.[11]

10.1.5 Self-Determination and Informed Consent

In general, the respect that we must show each other as people created in God's image would require that physicians take the time to inform their patients about both the preventive and curative steps necessary for their care so that they can make informed decisions. At the same time, physicians need not inform their patients of alternatives that are, in their estimation, medically futile. To this point the Jewish and American traditions agree.

There is, however, an important difference in degree in how the two traditions address these matters. American law puts great emphasis on patient autonomy, and physicians must therefore inform patients of absolutely every possible mishap for fear of being sued if the patient consented to the procedure without that knowledge. The Jewish tradition trusts physicians more than contemporary American law does; indeed, suits against physicians are virtually unheard of in the annals of Jewish law. Moreover, Jewish sources are concerned about the patient's mental health as much as his or her physical health. Consequently, the Jewish tradition would advise against physicians telling their patients absolutely everything that might go wrong in a procedure. Physicians must surely share information with their patients about dangerous outcomes that often occur with a given planned intervention, together with their probability; but when the likelihood of particular problems occurring is slight, maintaining the patient's good spirits would generally outweigh the need to provide information about unexpected outcomes.

---

[10] For a thorough discussion of these methodological issues, including why and how Judaism uses law to discern moral duties and the relationship of law to duties beyond the law, see Dorff (1998, pp. 395–417, 2007, chap. 6).

[11] For a more thorough treatment of these sources of Jewish morality, see Dorff (2003, pp. 311–344).

10.1.6 Truth-Telling and Confidentiality

Judaism strongly values telling the truth, and the Torah itself admonishes "stay far away from any lie" (Exodus 23:7).[12] At the same time, Judaism teaches that truth is not the only value, nor is it an absolute one. In hard cases, truth-telling must be weighed against other moral goods. So, for example, when telling the truth will only harm a person and not produce any other good, one must choose to remain silent or even gild the lily. A bride, then, is to be described on her wedding day as beautiful no matter how she looks, for tact in such circumstances takes precedence over truth (B. *Ketubbot* 16b-17a; S.A. *Even Ha'ezer* 65:1; cf. M.T. *Laws of Ethics* [*De'ot*] 7:1.) On the other hand, when writing a letter of recommendation for a job, the writer must reveal the applicant's weaknesses relevant to the job, for those may have a practical effect on the welfare of others.[13]

Similar guidelines apply to the caregiver-patient relationship. By and large, patients do better when they know what to expect; they feel infantalized and lose trust in their physicians and loved ones when relevant factors about their disease are hidden from them or misrepresented. In general, then, patients should be told the truth—calmly, clearly, and tactfully, to be sure, but the truth nonetheless. If the patient's disease is incurable, the patient should be told that that is so, but this should be accompanied by a description of what can be done physically to help the patient cope with his or her condition and how the patient's family, friends, rabbi, and other caregivers can help the patient emotionally and spiritually.

When, however, it is the judgment of the physician (and in the case of a child, the parents) that the patient would be better off not knowing, that is a reasonable choice; the patient's welfare takes precedence over the truth in such cases. Due care, though, must be given to deciding whether this is indeed such a case, for most often it is not.

10.2 Advanced Directives: Proxy Decisions and Living Wills

Jewish law would allow Jews to write an Advanced Directive nominating someone else to make medical decisions for them when they cannot do so themselves. The proxy, of course, would have no more authority in Jewish law to make medical decisions than the patient would have, and here it is important to remember that Jewish

[12] In context, that passage, like Exodus 20:13 in the Ten Commandments, may be talking specifically about the legal setting, warning that one not allege a false charge, but the later Jewish tradition understood it more broadly to forbid all falsehood. See, for example, B. *Ketubbot* 17a; B. *Shevuot* 30b, 31a; B. *Bava Mezia* 49a; M.T. *Laws of Ethics* (*De'ot*) 2:6; cf. 5:13. Moreover, other verses in the Bible itself, such as Psalms 101:7, Psalms 119:163, and Proverbs 13:5, condemn falsehood in general.

[13] For a more thorough discussion of Jewish norms governing language generally and in particular when truth is trumped by other considerations, see Dorff (2005, pp. 69–107, esp. 91–98). For the specific norms about giving references for schools or jobs, see Dorff and Gary (2014).

sources give the physician, as the medical expert caring for God's property, more authority relative to the patient or the surrogate than American law does. Still, Jews may appoint representatives to guide their health care.

In addition, Jews may fill out a living will to indicate how they would want decisions to be made in a variety of circumstances. In fact, all of the denominations of American Judaism have published such documents for the use of their constituents. Each reflects the particular denomination's understanding of the content and degree of authority of Jewish law. Orthodox Judaism asserts that Jewish law is binding, and so in the Orthodox Advance Directive (Rabbinical Council of America) patients declare that they are Orthodox Jews and therefore want their rabbi, who is identified and whose contact information is given, to make all medical decisions for them. Conservative Judaism also believes that Jewish law is binding but that it has changed in the past and should change in our own day to reflect modern circumstances and sensitivities, and so the Conservative Advance Directive (Mackler 1993) describes 17 common sets of questions that patients and their families must face in end-of-life decisions and gives them the options that can be justified in Jewish law but not those that cannot be so justified. Reform Judaism places great emphasis on the autonomy of individual Jews to decide for themselves how they are going to express their Jewish identity, and so the Reform advance directive (Address 1992, 2016) is a grid that has a number of possible medical interventions along the vertical axis, with the horizontal axis giving patients the options to choose among the following as they check off what they want: "I want," "I want treatment tried, but if no clear improvement, stop," "I am undecided," or "I do not want."

10.3 The Process of Death and Dying

10.3.1 General Concepts and Categories

When we consider issues at the end of life, a few definitions will set the stage for the discussion. *Murder* is the taking of another's life with malice aforethought and without a legal excuse (which include fighting in a justified war or killing in self-defense).[14] *Active euthanasia* is a positive act with the intention of taking another's life, but for benign purpose (for example, to relieve the person from agonizing and incurable pain). This can be voluntary—that the patient requests it—or involuntary (where the patient is incapable of expressing a desire, as, for example, when the patient is in a coma), but because it is always intended for the patient's wellbeing, it is often called "mercy killing." *Passive euthanasia* is a refusal to intervene in the

[14] The prohibition of murder, as in the sixth of the Ten Commandments, does not interdict all killing of humans. On the contrary, Judaism requires self-defense, even to the extent of killing one's attacker, for both individuals (Exodus 22:1; B. *Berakhot* 58a; B. *Yoma* 85b; B. *Sanhedrin* 72a) and communities (as in war) (Deuteronomy 20–21; M. *Sotah* 8:7 [44b]). See Dorff (2002, chap. 7).

process of a person's natural demise (withholding treatment) or withdrawing treatment so that nature can take its course.

Jewish sources prohibit murder in all circumstances, and they view all forms of active euthanasia as the equivalent of murder (M. *Semahot* 1:1–2; M. *Shabbat* 23:5 and B. *Shabbat* 151b; B. *Sanhedrin* 78a; M.T. *Laws of Murder* 2:7; S.A. *Yoreh De'ah* 339:2 and the comments of the Shakh and Rama there). This is true even if the patient asks to be killed. Because each person's body belongs to God, the patient does not have the right either to commit suicide or to enlist the aid of others in the act, and anybody who does aid in this plan commits murder. No human being has the right to destroy or even damage God's property.[15]

The patient does have the right, however, to pray to God to permit death to come,[16] for God, unlike human beings, has the right to destroy God's own property. Moreover, Judaism does permit passive euthanasia in specific circumstances, and in our day it is those circumstances that are of extreme medical interest.

Unfortunately traditional sources on this are sparse, for until the advent of antibiotics in 1938, physicians could do very little to impede the process of dying unless the problem could be cut out of the patient surgically—and even then, more often than not the patient would bleed out or die of infections. Because physicians can now do a great deal for the dying, Jews seeking moral guidance from the Jewish tradition must place a heavy legal burden on the few sources that reflect circumstances in the past in which people thought they had an effective choice of whether to delay death or not. This not only leads to considerable disagreement on specific, clinical issues; it also poses significant methodological questions as to how the tradition can be legitimately accessed and applied to contemporary circumstances so very different from the past.[17]

10.3.2 Determining Death

Classical Jewish sources use two criteria for death. One is the breath test, in which a feather is placed beneath the nostrils of the patient to see if it moves. The exegetical bases for this test are the verses in Genesis according to which "God breathes life into Adam" (2:6) and the Flood kills "all in whose nostrils is the breath of the spirit of life" (7:22) (B. *Yoma* 85a; *Pirkei de-Rabbi Eliezer*, ch. 52; *Yalkut Shim'oni*,

---

[15] This includes even inanimate property that belongs to us among human beings, for God is the ultimate owner. Cf. Deuteronomy 20:19; B. *Bava Kamma* 8:6, 7; B. *Bava Kamma* 92a, 93a; S.A. *Hoshen Mishpat* 420:1, 31.

[16] Cf. RaN, B. *Nedarim* 40a. The Talmud records such prayers: B. *Ketubbot* 104a, B. *Bava Mezia* 84a, and B. *Ta'anit* 23a. Note that this is not a form of passive euthanasia, for there people refrain from acting, but here God is asked to act.

[17] For a discussion of the methodological issues involved in deriving legal guidance from such stories, see the articles by David Ellenson, Louis Newman, Elliot Dorff, and Aaron Mackler in Dorff and Newman (1995, pp. 129–193). See also Dorff (2010).

"Lekh Lekha," no. 72), but there clearly is also a cogent, practical reason for using the breath test—namely, that it is easy to administer.

Later codifiers embraced the approach of insisting on both respiratory and cardiac manifestations of death. Some even held that the breath test is sanctioned by the Talmud only because it normally is a good indication of the existence of heartbeat, but actually it is the cessation of heartbeat that forms the core of the Jewish definition of death (Rashi on B. *Yoma* 85a; Rabbi Tzevi Ashkenazi, *Hakham Tzvi*, no. 77; Rabbi Moses Sofer, *Teshuvot Hatam Sofer, Yoreh De'ah*, no. 338). Moreover, in the sixteenth century Rabbi Moses Isserles ruled that "nowadays" we do not know how to distinguish accurately between death and a fainting spell or coma, and consequently even after the cessation of breath and heartbeat, we should wait a period of time before assuming that the person is dead (Isserles, S.A. *Yoreh De'ah* 338, gloss). Some contemporary rabbis claim that we should still wait 20 or 30 min after observing these signs, but others maintain that the accuracy of the sphygmomanometer and electrocardiogram on these matters permits us to revert to the traditional mode of defining death as cessation of breath and heartbeat (e.g., Bleich 1981, pp. 152–154).

Beginning with the Harvard criteria in 1968 for determining death on the basis of whole brain inactivity, as measured by a flatline reading on an electroencephalogram, as well as clinical tests subsequently adopted by all 50 American states and many other jurisdictions as the legal definition of death, most (but not all) Jewish authorities have also accepted "whole brain death," including the brain stem, as sufficient to determine death because that is the current indicator of death in contemporary medical practice (the primary reason given by Conservative rabbis like Goldfarb 1976 and Siegel 1976) or because when whole brain death occurs, no further breathing is possible, and so the traditional marker of cessation of breath is fulfilled (the primary reason given by Orthodox rabbis; see Steinberg 2003, pp. 695–711).[18] The Reform Movement officially adopted the Harvard criteria (presumably, as modified by the medical community) in 1980 (Jacob 1983, pp. 273–274). As a result, as described below, organ donation after brain death has been accepted by all Jewish movements and even seen as a positive commandment by the Conservative Rabbinate, and in 1988 the Orthodox Chief Rabbinate of the State of Israel even accepted brain death for purposes of heart transplants.

---

[18] For a summary of some of the varying Orthodox opinions up to 1978 in America, England, and Israel on defining the moment of death, see Goldman (1978, pp. 223–229). See also Rosner and Bleich (1979, pp. 367–371) and Bleich (1981, pp. 146–157). For the Israeli Chief Rabbinate's ruling, see Jakobovits (1989). For Conservative positions, see the opinion of Jack Segal, cited in Goldman (1978, pp. 229–230 n.42), Siegel (1975, 1976), and Goldfarb (1976). The first official endorsement of the new criteria for the Conservative Movement came in the approval of the Conservative Movement's Committee on Jewish Law and Standards in December, 1990 of the responsa by Rabbis Elliot N. Dorff and Avram Reisner (see Dorff 1991; Reisner 1991a), both of which assume and explicitly invoke the new medical definition.

10.3.3 Forgoing Life-Sustaining Treatment

When does the Jewish obligation to cure end, and when does the permission (or, according to some, the obligation) to let nature take its course begin?

Authorities differ. All agree that one may allow nature to take its course once the person becomes a *goses*, a moribund person. But when does that state begin? The most restrictive position is that of Rabbi J. David Bleich, who limits it to situations when all possible medical means are being used in an effort to save the patient and nevertheless the physicians predict that he or she will die within 72 h (1981, pp. 141–42). Others define the state of *goses* more flexibly (up to a year or more) or in terms of symptoms rather than time, and they then apply the permission to withhold or withdraw machines and medications more broadly (Jakobovits 1975, p. 124 and n.46; Reisner 1991a, especially pp. 56–62 [in Mackler 2000, pp. 245–53; online, pp. 18–25]).

In a rabbinic ruling approved by the Conservative Movement's Committee on Jewish Law and Standards, I noted that Jewish sources describe a *goses* as if the person were "a flickering candle," so that he or she may not even be moved for fear of inducing death (Dorff 1991, esp. pp. 19–26, see also 1998: chaps. 7–9, esp. pp. 198–202; in Mackler 2000, pp. 292–358, esp. 316–325; online pp. 83–91). This applies only to people within the last hours of life, for before then it is medically necessary to move patients so that their muscles do not wither and they do not develop infected bed sores. Consequently, I argued, the appropriate Jewish legal category to describe people with terminal, incurable diseases who may nevertheless live for months and even years is instead *terefah*. Permission to withhold or withdraw medications and machines would then apply to people as soon as they are in the state of being a *terefah*, that is, as soon as they are diagnosed with a terminal, incurable illness.[19]

One important operating principle in these matters is this: because Jewish law does not presume that human beings are omniscient, it is only the best judgment of the attending physicians that counts in these decisions. Even if some cure is just around the corner, we are not responsible for knowing that. We may and must proceed on the best knowledge available at the time and place at hand. If this means that the person is currently incurable, then machines and medications may be withdrawn and palliative care administered.

[19] Life itself, of course, is terminal, and so the restriction of this criterion to those states recognized as illnesses—and the very definition of illness or malady—becomes critical. A very good analysis of this has been suggested by Bernard Gert, Dan Clouser, and Charles M. Culver and discussed with clarity and applications by Ronald M. Green in Dorff and Zoloth (2015, pp. 257–273).

10.3.4 Artificial Nutrition and Hydration

In our own day, people in comas and those who cannot swallow are fed through tubes. All available forms of intubation are uncomfortable and pose some risk of infection, but they do give such patients the fluids and nutrients they need.

If the person has been in such a state for a number of months, however, and there seems to be little, if any, hope of recovery, may one remove such tubes? On one hand, just as all of us need food and liquids, the patient needs the artificial nutrition and hydration that flow through the tubes. Some (e.g., Bleich) therefore maintain that we must intubate and do anything possible to keep the patient alive, regardless of life quality, until he or she dies of other causes. Others (e.g., Rabbi Moshe Tendler) argue that one need not intubate, but once one has done so, one may not remove the artificial nutrition and hydration (Rosner and Tendler 1990, p. 54). Some (e.g., Reisner) specifically allow withholding or withdrawing medications and machines but require artificial nutrition and hydration (Reisner 1991a, pp. 62–64; in Mackler 2000, pp. 265–67; online, p. 25–27) until the patient dies of other causes.

On the other hand, the nutrients that enter the body through tubes look exactly like medications administered that way, and they lack many of the aspects of food—namely, differences in temperature, taste, and texture as well as ingestion by chewing and swallowing. Consequently, in the ruling I wrote for the Conservative Movement's Committee on Jewish Law and Standards, I ruled that although we must go through the motions of bringing in a normal food tray at regular meal times to a patient who cannot eat or drink normally, we need not administer nutrition and hydration artificially. We may do so, of course, and we should do so as long as there is a reasonable chance that the patient may recover. When that is no longer expected, however, so that the artificial nutrition and hydration are just prolonging the dying process, they may be removed (Dorff 1991, pp. 34–39, 1998, pp. 208–217; in Mackler 2000, pp. 348–54; online, pp. 100–106).

10.3.5 Curing the Patient, Not the Disease

The important thing to note, however, is that there is general agreement that a Jew need not use heroic measures to maintain his or her life but only those medicines and procedures that are commonly available in the person's time and place. We are, after all, commanded to *cure* based on the verse in Exodus 21:19, "and he shall surely cure him." We are not commanded to sustain life *per se*.[20] Thus, on the one

[20] Thus the Talmud specifically says, "We do not worry about mere hours of life" (B. *Avodah Zarah* 27b). The Talmud also says, however, that we may desecrate the Sabbath even if the chances are that it will only save mere hours of life (B. *Yoma* 85a). The latter source has led some Orthodox rabbis to insist in medical situations that every moment of life is holy and that therefore every medical therapy must be used to save even moments of life; see, for example, Bleich (1981, pp. 118–119, 134–145). The only exception is when a person is a *goses*, which Rabbi Bleich

hand, as long as there is some hope of cure, heroic measures and untested drugs *may* be employed, even though they come with an elevated level of risk. On the other hand, though, physicians, patients, and families who are making such critical care decisions are *not* duty-bound by Jewish law to invoke such therapies.

This should help us deal with a common phenomenon. A person is suffering from multiple, incurable illnesses, one of which is bound to cause death soon. It often happens that such a person develops pneumonia, and doctors are then in a quandary. A generally healthy person who contracts pneumonia would be treated with antibiotics, and often the drugs would bring cure. In those situations, according to Jewish law, both the physician and the patient would be required to use antibiotics, and few would need Jewish law to convince them to do so. But what happens in the case described above? The physician can probably cure the pneumonia, but that would only restore the patient to the pain and suffering caused by his or her other terminal maladies. The alternative would be to let the patient die of the pneumonia so that death would come more quickly.

From the perspective of Jewish law, the question is whether our inability totally to cure the person gives us the right to refrain from curing what we can. Normally we do not have this right. So, for example, we must try to cure the pneumonia of a child who has Down's Syndrome, even though we cannot cure the Down's Syndrome. If a person has a terminal illness, however, we would not need to intervene; we may rather let nature take its course. We must view the person as a whole rather than consider each individual disease separately. Therefore even though we could probably cure the pneumonia, and even though the means for doing so are not unusual at all, nevertheless *the person* cannot be cured, and therefore we may refrain from treating the pneumonia if that will enable the patient to die less painfully. This is in line with the strain in Jewish law that does not automatically and mechanically assume that preservation of life supersedes all other considerations, but rather judges according to the best interests of the patient.[21]

---

defines as within 72 h of death, at which time passive, but not active, euthanasia may be practiced. He then uses the source in *Avodah Zarah* to permit removing only hazardous therapies that may hasten death if they do not succeed in lengthening life. Rabbi Bleich's position is *not*, however, necessitated by the sources. On the contrary, they specifically allow us (or, on some readings, command us) not to inhibit the process of dying when we can no longer cure, even long before 72 h before death (however that is predicted).

[21] Tosafot, B. *Avodah Zarah* 27b, s.v., *lehayyei sha'ah lo hyyshenan*. See Dorff (1991, pp. 15–17, 43, n.22) (in Mackler 2000, pp. 311–114; online, pp. 79–82), (1998, pp. 202–208). For a contrasting interpretation of this source, see Reisner (1991a, pp. 56–57, 72, n.21) (in Mackler 2000, pp. 245–247, 255–257, n.22; online, pp. 18–19, pp. 26–38 n.22).

10.3.6 Pain Control and Palliative Care

The fact that Jewish law does not require the use of heroic measures means that a Jew may enroll in a hospice program in good conscience and that rabbis may suggest this in equally good conscience. There are some buildings called "hospices," but "hospice" care typically does not take place in a special facility. Rather the patient lives at home as long as possible, doing whatever he or she can do. The word *hospice* thus designates not so much a building, but a form of care. The goal of hospice care is not to cure the disease, but to make the patient as comfortable as possible. Thus the patient is still fulfilling the mandate of Jewish law to gain the aid of doctors when ill, but the goal of such help is now no longer cure, which has been deemed impossible, but comfort. In seeking to accomplish that goal, it is permissible to prescribe high doses of pain medication despite their potential of becoming addictive. Some (e.g., Rabbi Reisner) would allow only amounts that physicians are confident will not bring about the patient's death; I, though, would allow even amounts that may actually hasten the patient's death, as long as the intent is not to kill the person but rather to alleviate his or her pain.[22]

Hospice care, though, also crucially includes all the non-medical ways in which people are supported when they go through crises. Family and friends provide the psycho-social care that is so crucial to everyone, sick or not, as they keep the patient company and make the patient feel that he or she is still part of their world and not simply a locus of illness and pain. Nurses, social workers, and rabbis may also be involved at various points in the patient's care.

Moreover, if the doctors can use extraordinary means but only at great cost or by inflicting great pain and even then with only a slight possibility of cure, Jewish law would permit such action but would not require it. Consequently, a Jew may legitimately refuse supererogatory medical ministrations and may sign an Advance Directive for health care that indicates his or her desire to decline such care, choosing instead only to alleviate whatever pain is involved in dying. When cure is not possible, both the patient and the physician cease to have an obligation to do more medically than ease pain. Similarly, family and friends should not pressure the patient or physician to employ extraordinary or futile measures; they should instead focus on their continuing duties to visit the sick and provide all forms of physical and spiritual comfort. Although the various forms of Advance Directives that all four movements in American Judaism have produced differ in tone and substance on a number of matters, they all permit hospice care.

[22] In other words, Rabbi Reisner does not accept the "double effect" argument, while I do. See Reisner (1991a, pp. 66, 83–85, ns.50–52) (in Mackler 2000, pp. 269–270, pp. 283–286, ns.12–14; online, pp. 29, 55–57 ns.50–52) and see, in contrast, Dorff (1991, pp. 17–19, 43–45, ns.24–27) (in Mackler 2000, pp. 314–316, 328–330, ns.7–10; online pp. 82–83, 113–115 ns.24–27), (1998, pp. 185–186, 218–219, 379 n.76). See also Rabbi Reisner's summary of the differences between the Dorff and Reisner positions, Reisner (1991b).

10.3.7 Medical Experimentation/Research

Because of the strong imperative within the Jewish tradition to heal, medical experimentation is not only condoned but prized, and the artificial nature of the cures that researchers might concoct was never an issue. That is, in contrast to natural law perspectives that see natural means to address medical questions as superior to those made by humans, Judaism sees both as permissible, for human beings and their artifacts are after all part of the world we inhabit. What is required is that the new therapy offer the hope that the person will be helped in a way or to an extent that any less dangerous therapy does not, and that due experimentation on animals or cell lines be conducted before the new therapy is tried on human beings. If the patient—whether adult or child—suffers from an incurable illness, experimental procedures or drugs may be used in an attempt to cure the illness, even if they pose the risk of hastening the person's death or fail to effect a cure. The intention of all concerned, though, must be to try to heal the person and not to commit active euthanasia.

If a fetus has been aborted for reasons approved by Jewish law, it may be used for purposes of transplantation or research. If a family member suffers from leukemia and no appropriate bone marrow match is available, a married couple may seek to have another child in an attempt to find such a match, but only if they will not abort the child even if it becomes clear that the child is not the match they seek. They may also choose to have a child through *in vitro* fertilization so that they can choose an embryo that will be a match. Clearly, they must then treat the new child not simply as a means to an end but as their full-fledged child.

A person may volunteer to undergo an experimental procedure that holds out no hope to improve his or her own health but may increase medical knowledge only if it subjects the person to minimal or no risk. One's duty to preserve one's own life takes precedence over one's obligation to help other people preserve theirs.

10.3.8 Social Support of the Sick

Caring for a sick or dying person is not a matter of physical ministrations alone. The Jewish tradition was well aware that recovery is often dependent upon the social and psychological support—or lack thereof—that family and friends provide. Indeed, in cases where people ask to die, it is often because nobody is around to pay any attention to them.

To combat this, the Jewish tradition imposes the obligation on us of *biqqur holim*, visiting the sick. This is a *mitzvah*, a commanded act and an expected behavior, not only for rabbis, but for all Jews. Many synagogues and Jewish social groups thus have an active *biqqur holim* committee who take upon themselves the duty of ensuring that their sick and dying members are regularly visited. Rabbis, psychologists, and social workers sometimes train the members of the society on how to visit a bedridden person. This includes simple techniques like not standing over the bed but rather sitting down next to the patient so that the two of you are on the same plane. It also includes more complex matters, like how to engage the patient in conversation about matters beyond the food served for lunch that day.[23]

The Jewish tradition, then, not only obligates us to cure but also to care. Our medical facilities and our residence homes need to be not only medically sound, but warm, caring places. Moreover, our communities must consist of caring people who know that the Torah is serious when it says, "Love your neighbor as yourself" (Leviticus 19:18).

10.4 Care of the Deceased

10.4.1 General Principles

The treatment of the deceased in Jewish law depends on two primary principles. The general tenet that governs treatment of the body after death is *kevod ha'met*, that is, we should honor the dead body. This is not only demanded by respect for the deceased person; it also derives from the theological tenet that the body, even in death, remains God's property.

The other principle that affects the topics of this section is that of *pikkuaḥ nefesh*, the obligation to save people's lives. This tenet is so deeply embedded in Jewish law that, according to the Rabbis, it takes precedence over all other commandments except the prohibitions of murder, idolatry, and incestuous or adulterous sexual intercourse (B. *Sanhedrin* 74a-b). (That is, if, for example, one's choice is to murder someone else or give up one's own life, one must give up one's own life. If, however, one needed to violate the Sabbath laws or steal something to save one's own life or someone else's, then one is not only permitted, but commanded to violate the laws in question to save a life.)

Jews are commanded not only to do virtually anything necessary to save their own lives; they are also bound by the positive obligation to take steps to save the lives of others. The imperative to do so is derived from the biblical command, "Do not stand idly by the blood of your neighbor" (Leviticus 19:16). This means, for example, that if you see someone drowning, you may not ignore him or her but must do what you can to save that person's life (B. *Sanhedrin* 73a). That is, in contrast to the law in most American states (Volokh 2009), Jewish law imposes a duty to rescue.

What happens, though, when you can only save your life or someone else's? Whose life takes precedence?

The opinion that ultimately wins the day in Jewish legal literature is that of Rabbi Akiba (B. *Bava Mezia* 62a). Under morally impossible circumstances, when an

[23] For a description of how Jewish tradition bids us to act when visiting the sick, see Dorff (1998, pp. 255–264, 2005, pp. 157–162).

untoward result will happen no matter what one does, Rabbi Akiba directs us to remain passive and let nature take its course.

When these underlying principles in mind, we are now prepared to address a number of clinical issues concerning the corpse.

10.4.2 Cremation

Jewish law prohibits cremation as the ultimate form of dishonor of the dead. Cremation also represents the active destruction of God's property, and it is improper for that reason as well. In the generations after Hitler's gas chambers, burning the bodies of our own deceased seems especially inappropriate, given that it is treating the bodies of our loved ones as Hitler did to our ancestors. Cremations obviate the need to use land for burial, but the process of cremation is not green: it includes substantial uses of energy to burn the body, and most of the burned body goes up in smoke, thus polluting the air (Calgary Cooperative Memorial Society). Significant limitations on land for burial, especially in large cities, have, however, led to rabbinic rulings about alternative methods of burial, including two or more people buried on top of each other with a specified amount of earth between each casket, reburial of the bones in a smaller plot after disintegration of the body in order to free the larger plot for another burial, burial in a mausoleum, and alkaline hydrolysis (Kalmanofsky 2017).

10.4.3 Autopsies

The two procedures that are permitted to interrupt the normal Jewish burial process are autopsies and organ transplants. Autopsies were known in the ancient world, but Jewish sources indicate that they were largely looked upon as violations of human dignity. As the prospects of gaining medical knowledge from autopsies has improved, however, many rabbis have come to view them more favorably. A definitive position was enunciated by Israeli Chief Rabbi Isaac Herzog in his 1949 agreement with Hadassah Hospital. Under this agreement, autopsies are sanctioned only when one of the following conditions obtains:

(a) The autopsy is legally required.
(b) In the opinion of three physicians, the cause of death cannot otherwise be ascertained.
(c) Three physicians attest that the autopsy might help save the lives of others suffering from an illness similar to that from which the patient died.
(d) A hereditary illness was involved, and performing the autopsy might safeguard surviving relatives.

In each case, those who perform the autopsy must do so with due reverence for the dead, and upon completion of the autopsy, they must deliver the corpse and all of its parts to the burial society for interment. This agreement was incorporated into Israeli law 4 years later.

Whether an autopsy is justified for legal or medical reasons, it is construed not as a dishonor of the body but, on the contrary, as an honorable use of the body to help the living. New procedures, such as a needle biopsy of a palpable mass or a peritoneoscopy with biopsy may soon accomplish most of the same medical objectives as autopsies have done without invading the corpse to the same degree, and that would clearly be preferable from a Jewish point of view.[24]

10.4.4 Organ and Tissue Transplantation

When considering organ transplantation from a dead person, the overriding principles of honoring the dead (*kevod ha-met*) and saving people's lives (*pikkuah nefesh*) work in tandem. That is, saving a person's life is so sacred a value in Judaism that if a person's organ can be used to save someone else's life, it is actually an honor to the deceased person to use the organ in that way. This is certainly the case if the person completed an advanced directive, either orally or in writing, indicating willingness to have portions of his or her body transplanted; but even if not, the default assumption in Jewish law, in contrast to American law, is that a person would be honored to donate his or her organs to help someone else live.

There are, however, some restrictions on this. Rabbis have differed on the circumstances under which organs may be transplanted. The most restrictive opinion would limit donations to cases in which there is a specific patient before us (*holeh shelefaneinu*) and that person's life or an entire physical faculty of the person is at stake. So, for example, if the person can see out of one eye, a cornea may not be removed from a dead person, according to this opinion, to restore vision in the other eye. Only if both eyes are failing, such that the potential recipient would lose all vision and therefore incur increased danger to life and limb, may a transplant be performed. Moreover, the patient for whom the organ is intended must be known and present; donation to organ banks is not permitted (Bleich 1981, pp. 129–133 [esp. 132], 162–168 [esp. 166–67]).

This is definitely an extreme position. Most rabbis, including Orthodox ones, would expand both the eligibility of potential recipients and the causes for which an organ may be taken. By graphically indicating that the person has died, burial helps people gain the emotional catharsis and closure that they need. Therefore the permission of the family is necessary not only to accord with American law, but also to assure that even without burial of all of a person's body, relatives of the deceased in

[24] See Jakobovits (1975, pp. 150, 278–283, and, more generally, pp. 132–152). The Chief Rabbinate's ruling and the Israeli Anatomy and Pathology Act of 1953 are cited at p. 150, and Rabbi Jakobovits' own opinion can be found on pp. 278–283.

this particular case can effectively carry out the mourning process so that they can have psychological closure and return to their lives in full. If this is not possible, families may refuse to give permission for organ donation or dissection, and they should not be made to feel guilty for not doing so.

With family agreement, though, most rabbis would permit the transplantation of a cornea into a person with vision in only one eye on the grounds that impaired vision poses enough of a risk to the potential recipient to justify the surgical intrusion of the corpse necessary to provide the cornea. Some would not require that the person be nearby and ready for transplantation but only identified. In these days of organ banks, however, most rabbis would be satisfied that there is sufficient demand for the organ such that it is known that it will eventually, but definitely, be used for purposes of transplantation. So, for example, the Rabbinical Assembly, the organization of Conservative rabbis, approved a resolution in 1990 to "encourage all Jews to become enrolled as organ and tissue donors by signing and carrying cards or drivers' licenses attesting to their commitment of such organs and tissues upon their deaths to those in need" (*Proceedings* 1990, p. 279).[25]

According to the rulings of the vast majority of rabbis who have written on this, then, cadaveric donations of any bodily part would be permissible, including even vital organs like the heart. Indeed, donating your organs upon your death is considered an act of special kindness (*ḥesed*), for using the body to enable someone else to live with full human faculties is not a desecration of the body but rather a consecration of it. At its meeting in December 1995, the Conservative Movement's Committee on Jewish Law and Standards went even further, approving a ruling by Rabbi Joseph Prouser making cadaveric organ donation not only an act of special kindness (*ḥesed*), but a positive obligation (*ḥovah*). The chief considerations that motivated Rabbi Prouser and the committee to take this position were the lives that can be saved through organ transplantation and the importance of ensuring that living relatives not be pressured into making risky donations of their own organs due to a shortage of organs from cadavers (Prouser 1995).

[25] Although somewhat dated, a good summary of the positions of all three movements on organ transplantation, with relevant quotations from responsa and other official position statements, can be found in Goldman (1978, pp. 211–237). This includes quotations from two earlier responsa approved by the Conservative Movement's Committee on Jewish Law and Standards. A similar stance can be found in the work of two other Conservative rabbis, namely, Klein (1975, chap. 5), and Feldman (1986, pp. 103–108).

For a summary of Orthodox positions, see Jakobovits (1975, pp. 278–291), Rosner and Bleich (1979, pp. 387–400), and more recently, Steinberg (2003, pp. 695–711).

For a Reform position on this, see Freehof (1956, 1968 [both reprinted in Jacob 1983, pp. 288–296], 1960, pp. 130–131, 1974, pp. 216–233). In a March, 1986 responsum, the Central Conference of American Rabbis as a body officially affirmed the practice of organ donation, and the synagogue arm of the Reform Movement, through its Committee on The Synagogue as a Caring Community and Bio-Medical Ethics, published a manual for preparing for death that specifically includes provision for donation of one's entire body or of particular organs to a specified person, hospital, or organ bank for transplantation or for research, medical education, therapy of another person, or any purpose authorized by law. The manual is Address (1992).

The only concern is to make sure that the donor is indeed dead before the donation takes place. For most rabbis, including the Chief Rabbinate of the State of Israel, that means, in the age of modern technology, the cessation of all brain wave activity; for a few, it still requires cessation of breath and heartbeat. In recent times, some want to return to the old criteria to justify "non-heart beating donors" even when some brain wave activity is detectable (Sharzer 2010) while others worry that this will all too easily motivate physicians to curtail the treatment of the donor. At this stage, rabbinic opinion on this new procedure has not been settled. Whichever definition of death is used, though, once death has occurred, the prohibitions against desecration of the dead, deriving benefit from the dead, and delaying the burial of the dead are suspended for the greater good of saving the recipient's life or restoring his or her health, thus giving even greater honor to the deceased (Feldman and Rosner 1984, pp. 67–71; Goldman 1978, pp. 211–237).

10.4.5 Donating One's Body to Science

May one donate one's entire body to science for purposes of dissection by a medical student as part of his or her medical education? Objections to this center around the desecration of the body involved in tearing it apart and the delay in its burial until after the dissection. Even so, Israeli Chief Rabbi Herzog issued the following statement in 1949:

> The Plenary Council of the Chief Rabbinate of Israel...does not object to the use of bodies of persons who gave their consent in writing of their own free will during their lifetime for anatomical dissections as required for medical studies, provided the dissected parts are carefully preserved so as to be eventually buried with due respect according to Jewish law. (Quoted in Jakobovits 1975, p. 150)

Although a few medical schools are now experimenting with teaching anatomy through computer simulation programs, dissection remains the standard way of teaching anatomy first-hand to first-year medical students. Consequently, participating in medical education in this way is an honor to the deceased and a real kindness in that it helps the living. The levity that sometimes accompanies dissection is not because medical students find dissection funny; it is rather a way for them to dissipate their discomfort in handling a corpse. No disrespect is intended, and therefore dissection is not objectionable on that ground.

Rabbi Isaac Klein cites yet another argument to permit the donation of one's body to science:

> In a country where the Jews enjoy freedom, if the rabbis should refuse to allow the Jewish dead to be used for medical study, their action will result in *hillul ha-shem* [a desecration of God's Name], for it will be said that the Jews are not interested in saving lives; there is (therefore) reason to permit it. (Klein 1975, p. 41)

The only restrictions imposed by those who permit people to donate their body to science are that the remains ultimately be buried according to Jewish law and custom and, for the reasons described above, that the family of the deceased agrees.[26]

Permission for donating one's body to science would not apply, however, if there are ample bodies available for dissection. Then there would be no special gift being given by the donor to future physicians and their patients, and there would be no particular taint involved if Jews do not generally donate their bodies for this purpose. Therefore, because medical schools currently have more than enough bodies from county morgues, largely bodies of unknown people that have been abandoned, Jews need not and therefore should not offer to have their bodies dissected, for there is no medical necessity to set aside the honor due a corpse according to the Jewish concept of *kevod ha-met*.

10.5 Epilogue

In sum, the Jewish tradition values the life of the disabled and the dying just as much as it does the life of the fully abled. Through its legal rulings, theology, stories, and proverbs, it seeks to alleviate suffering, to save lives and bodily functions, and to give meaning to the full breadth of life, while yet recognizing human mortality and seeking to help people cope with dying both physically and spiritually. It is fitting, then, that we end with the Psalmist's instruction to us to make every day count:

> The span of our life is seventy years,
> or, given the strength, eighty years....
> They pass by speedily, and we are in darkness....
> Teach us to count our days rightly,
> that we may obtain a wise heart...
> May the favor of the Lord, our God, be upon us,
> let the work of our hands prosper,
> O prosper the work of our hands!
> (Psalms 90:10, 12, 17)

[26] Permission of the donor or his family must be procured so that the transplant does not constitute a theft according to Chief Rabbi Unterman's responsum in Goldman (1978, p. 226). Feldman and Rosner (1984, p. 68) say that the family's permission is only advisable in Jewish law, but it is mandatory in American law; that, however, would make it religiously required of American Jews as well under the Jewish legal principle of "the law of the land is the law" (B. *Nedarim* 28a; B. *Gittin* 10b; B. *Bava Kamma* 113a; B. *Bava Batra* 54b-55a). This principle is described in Dorff and Rosett (1988, pp. 515–523). See also Klein (1975, pp. 40–41).

References

Address, Richard F., ed. 1992. *A time to prepare: A practical guide for individuals and families in determining one's wishes for extraordinary medical treatment and financial arrangements*. Philadelphia: Union of American Hebrew Congregations Committee on Bio-Medical Ethics. Updated version, 2016: http://jewishsacredaging.com/wp-content/uploads/2016/05/Expanded-A-Time-To-Prepare-Jewish-Sacred-Aging.pdf.

Bleich, J. David. 1977. Smoking. *Tradition* 16/4 (Summer): 130–133. reprinted in Dorff and Rosett 1988, pp. 349–352.

———. 1981. *Judaism and healing*. New York: Ktav.

Calgary Cooperative Memorial Society. *More on cremation and its impact on the environment*. https://www.calgarymemorial.com/effect-of-cremation-on-environment.html.

Dorff, Elliot N. 1991. A Jewish approach to end-stage medical care. *Conservative Judaism* 43/3: 3–51. reprinted in Mackler 2000, pp. 292–358 and, in an updated form, in Dorff 1998, chaps. 7–9. Also available in its original form online at https://www.rabbinicalassembly.org/sites/default/files/assets/public/halakhah/teshuvot/19861990/dorff_care.pdf.

———. 1998. *Matters of life and death: A Jewish approach to modern medical ethics*. Philadelphia: Jewish Publication Society.

———. 2002. *To do the right and the good: A Jewish approach to modern social ethics*. Philadelphia: Jewish Publication Society.

———. 2003. *Love your neighbor and yourself: A Jewish approach to modern personal ethics*. Philadelphia: Jewish Publication Society.

———. 2005. *The way into tikkun olam (repairing the world)*. Woodstock: Jewish Lights.

———. 2007. *For the love of God and people: A philosophy of Jewish law*. Philadelphia: Jewish Publication Society.

———. 2010. Applying Jewish law to new circumstances. In *Teferet Leyisrael: Jubilee volume in honor of Israel Francus*, ed. Joel Roth, Menahem Schmelzer, and Yaacov Francus, 189–199. New York: Jewish Theological Seminary.

Dorff, Elliot N., and Marc Gary. 2014. *Providing references for schools or jobs*. https://www.rabbinicalassembly.org/sites/default/files/assets/public/halakhah/teshuvot/2011-2020/providingreferences.pdf.

Dorff, Elliot N., and Louis Newman. 1995. *Contemporary Jewish ethics and morality: A reader*. New York: Oxford.

Dorff, Elliot N., and Arthur Rosett. 1988. *A living tree: The roots and growth of Jewish law*. Albany: State University of New York Press.

Dorff, Elliot N., and Laurie Zoloth, eds. 2015. *Jews and genes: The genetic future in contemporary Jewish thought*. Philadelphia: Jewish Publication Society.

Eisenstein, J.D., ed. 1915. *Ozar Midrashim*. Vol. II. New York: Eisenstein.

Feldman, David M. 1986. *Health and medicine in the Jewish tradition*. New York: Crossroad.

Feldman, David M., and Fred Rosner, eds. 1984. *Compendium on medical ethics*. New York: Federation of Jewish Philanthropies of New York.

Freehof, Solomon B. 1956. The use of the cornea of the dead. *C.C.A.R. [Central Conference of American Rabbis] Yearbook* 66: 104–107.

———. 1960. *Reform response*. Cincinnati: Hebrew Union College Press.

———. 1968. Surgical transplants. *C.C.A.R. Yearbook* 78: 118–121.

———. 1974. *Contemporary reform responsa*. Cincinnati: Hebrew Union College Press.

———. 1977. *Reform responsa for our time*. Cincinnati: Hebrew Union College Press.

Goldfarb, Daniel C. 1976. The definition of death. *Conservative Judaism* 30/2: 10–22.

Goldman, Alex J. 1978. *Judaism confronts contemporary issues*. New York: Shengold Publishers.

Jacob, Walter, ed. 1983. *American reform responsa*. New York: Central Conference of American Rabbis.

Jakobovits, Immanuel. 1975. *Jewish medical ethics*. New York: Bloch.

Jakobovits, Yoel. 1989. [Brain death and] heart transplants: The [Israeli] Chief Rabbinate's directives. *Tradition* 24 (4): 1 14.
Kalmanofsky, Jeremy. 2017. *Alternative Kevura methods*. https://www.rabbinicalassembly.org/sites/default/files/public/halakhah/teshuvot/2011-2020/alternative-burial_kalmanofksy_new-version.pdf.
Klein, Isaac. 1975. *Responsa and Halakhic studies*. New York: KTAV.
Mackler, Aaron L. 1993. *Jewish medical directives for health care*. https://www.rabbinicalassembly.org/sites/default/files/assets/public/halakhah/teshuvot/19861990/mackler_care.pdf.
Mackler, Aaron. 2000. *Life and death responsibilities in Jewish biomedical ethics*. New York: The Jewish Theological Seminary of America (Finkelstein Institute.
Nahmanides. 1963. *Kitvei Haramban*, ed. Charles Chavel Jerusalem: Mosad Harav Kook.
Proceedings of the [Conservative] *Rabbinical Assembly*. 1983. Vol. 44. New York: Rabbinical Assembly.
Proceedings of the [Conservative] *Rabbinical Assembly*. 1990. Vol. 52. New York: Rabbinical Assembly.
Prouser, Joseph H. 1995. *Hesed or Hiyyuv? The obligation to preserve life and the question of post-mortem organ donation*. https://www.rabbinicalassembly.org/sites/default/files/assets/public/halakhah/teshuvot/19912000/prouser_chesed.pdf.
Rabbinical Council of America. n.d. *Halachic health care proxy*. http://www.rabbis.org/pdfs/hcp.pdf.
Reisner, Avram Israel. 1991a. A Halakhic ethic of care for the terminally ill. *Conservative Judaism* 43/3: 52–89. reprinted in Mackler 2000, pp. 239–291; and available online at https://www.rabbinicalassembly.org/sites/default/files/assets/public/halakhah/teshuvot/19861990/reisner_care.pdf.
———. 1991b. Mai Beinaiyhu [What is the difference between them?]. *Conservative Judaism* 43/3: 90–91. reprinted in Mackler 2000, pp. 236–238; and available online at https://www.rabbinicalassembly.org/sites/default/files/assets/public/halakhah/teshuvot/19861990/maibeinaihu.pdf.
Rosner, Fred, and J. David Bleich, eds. 1979. *Jewish bioethics*. New York: Sanhedrin Press.
Rosner, Fred, and Harav Moshe Tendler. 1990. *Practical medical halachah*. 3rd ed. Hoboken: Ktav.
Sharzer, Leonard. 2010. *Organ donation after cardiac death*. https://www.rabbinicalassembly.org/sites/default/files/assets/public/halakhah/teshuvot/20052010/dcd_teshuvah_final%2002-10.pdf.
Siegel, Seymour. 1975. Fetal experimentation. *Conservative Judaism* 29/4: 39–48.
———. 1976. Updating the criteria of death. *Conservative Judaism* 30/2: 23–30.
Steinberg, Avraham. 2003. *Encyclopedia of Jewish medical ethics*. 3 vols. Trans. Fred Rosner. Jerusalem/New York: Feldheim.
Volokh, Eugene. 2009. *Duty to rescue/report statutes*. http://volokh.com/2009/11/03/duty-to-rescuereport-statutes/.

Chapter 11
Buddhism and Brain Death: Classical Teachings and Contemporary Perspectives

Damien Keown

Abstract This chapter discusses the concept of brain death and its implications for medical practice as seen from a Buddhist perspective. After a brief introduction to Buddhist teachings, the chapter considers the compatibility of brain death with the classical Buddhist understanding of death as found in the earliest sources. Certain conceptual discrepancies are highlighted which problematize the removal of vital organs from patients who may be judged to be still alive on Buddhist criteria. The implications of this for contemporary medical practice are then explored in two Asian countries, Japan and Thailand. While different issues arise in each case, in neither country does the concept of brain death appear to have been well received at a popular level, a situation aggravated by medical scandals surrounding organ transplantation.

11.1 Introduction

I will begin with a brief summary of Buddhism for those unfamiliar with the tradition. Buddhism was founded in the fifth century BCE by an individual from an influential family in north-east India who experienced a spiritual awakening at the age of 35 and spent the remaining 45 years of his life as an itinerant teacher. This individual, whom we know as the Buddha, meaning "awakened one," established a monastic order called the Sangha which was instrumental in spreading his teachings throughout Asia. These teachings are called the Dharma, and collectively these three items—the Buddha, the Dharma, and the Sangha—are known as the "three jewels" of Buddhism.

Buddhist teachings are summed up in a formula known as the Four Noble Truths. The first noble truth states that human existence is difficult and often painful, involving, as Buddhists believe, a potentially endless cycle of rebirth in which individuals are constantly exposed to suffering. The second truth locates the root of the problem

D. Keown (✉)
History Department, University of London Goldsmiths, London, UK
e-mail: d.keown@gold.ac.uk

T. D. Knepper et al. (eds.), *Death and Dying*, Comparative Philosophy of Religion 2, https://doi.org/10.1007/978-3-030-19300-3_11

just described in ignorance of what causes this suffering and emotional attachment to things that cannot fulfil us. The third teaches that there is a state free from suffering known as nirvana, and the fourth sets out a path that leads to this state. This path calls for a balanced program of living that eschews extremes and emphasizes virtuous conduct, meditation, and wisdom.

As Buddhism developed, two main traditions emerged. The earlier and more conservative is Theravāda Buddhism, which is found in south and southeast Asian countries like Sri Lanka and Thailand. A second, more broadly-based movement known as Mahāyāna Buddhism, developed around the beginning of the Christian era and spread to places like Tibet, China, and Japan. Further subdivisions occurred and the religion has never had a central administration or supreme authority.

This fragmentation problematizes to some degree one of the objectives of the present volume, since it hard to speak about challenges to orthodoxy in the context of such diversity.[1] Nor will we will find much evidence of tensions between tradition and modernity within Buddhism itself. The tensions we will encounter, rather, are between classical Buddhist teachings and modern developments such as the medicalization of death and the legalization of a new concept of death.

I will endeavor to explain how these tensions arise and explore their consequences in two culturally distinct parts of the Buddhist world, Japan and Thailand.[2] The reason for selecting these countries is twofold: first, they provide examples of each of the main families of Buddhism: Mahāyāna Buddhism in Japan, and Theravāda Buddhism in Thailand. And second, Japan and Thailand are countries that have both recognized brain death, and where cadaver organ transplants are currently performed. While differences between them will become apparent, I think we will also discern similarities stemming from their common Buddhist heritage.

The focus of our discussion will be the concept of brain death and its implications for medical practice. I first addressed the subject over twenty years ago in a discussion of end-of-life issues in my book *Buddhism and Bioethics*. At that time, I expressed the view that the concept of brain death would be acceptable to Buddhism, and that brain death was identical with human death. Since then I have come to doubt this assessment and now believe that although brain death usually heralds the imminent demise of the patient, it does not equate to death itself.[3] A survey of "intensivists" (medical staff who work in intensive care) carried out by Margaret Lock suggests they share a similar view. She writes "Among the thirty-two physician intensivists interviewed, not one thinks brain death signals the end of biological life, although everyone agreed brain death will lead to complete biological death" (2002, p. 243).[4] While brain dead patients may be dying, in other words, they are not yet dead, at least on an everyday understanding of what death means.

---

[1] For reflections on what might constitute a "Buddhist view," see Keown (2001, p. 12ff).

[2] For a Tibetan Buddhist perspective on the questions discussed here see Tsomo (2006).

[3] I set out my reasons for this in a subsequent paper (Keown 2010) parts of which I draw on here.

[4] In connection with these interviews Lock notes, "Tellingly, among the 32 doctors interviewed, only six have signed their donor cards …. When I pressed for reasons, no one gave me very convincing answers" (2002, p. 249).

If the above is correct, Buddhists face a conflict arising from the motivation to help others, through the donation of organs,[5] and respect for the moral principle of *ahiṃsā*, or non-harming. This principle is enshrined in the first of Buddhism's five moral precepts which enjoins us to do no harm. The central concern is that if brain-dead patients are not really dead, to practice solid organ explanation—such as the removal of the heart or other vital organs—will itself cause the somatic death of the patient. The fact that this is done in order to save lives may be a mitigating factor, but in Buddhist terms it still constitutes the intentional killing of a living human being.

The structure of the present chapter is as follows. I will first discuss the concept of brain death in part one, and then in part two present the classical Buddhist understanding of death as found in the earliest sources. In part three, I review contemporary aspects of the brain death question in Japan and Thailand.

11.2 The Concept of Brain Death

The universally recognized sign of death has always been bodily putrefaction. From the beginning of the nineteenth century, it became common for medical practitioners to locate the onset of this process at the point when the heart and lungs stop functioning. Subsequently, in 1968, a new definition of death as "irreversible coma" was proposed by a committee of the Harvard Medical School. This proposal emerged not as the result of disinterested reflection on human mortality but as a solution to two pressing problems. The first was that of brain-damaged patients being kept alive by machines. The Harvard report spoke of the "burden" imposed by such a condition on the patients themselves, and "on their families, on the hospitals, and in those in need of hospital beds already occupied by these comatose patients." As we shall see, this problem is particularly acute in Thailand. The second problem was that "[o]bsolete criteria for the definition of death can lead to controversy in obtaining organs for transplantation" (quoted in Lock 2002, p. 89).

The new criterion solved both problems at a stroke by legitimizing the withdrawal of life support from comatose patients, and allowing an individual to be declared dead for the purposes of organ transplants before the heart had stopped beating. Given its advantages, medical bodies around the world quickly accepted the new standard, and the concept of irreversible coma—or "brain death" as it became known—has since been enshrined in the legislation of many nations in a variety of formulations and protocols.[6] In the USA it was incorporated into the

[5] The Buddhist word for generosity (*dāna*) is etymologically related to the English "donor." *Dāna* is the first of the six perfections of a bodhisattva.

[6] The conclusion of a worldwide survey published in *Neurology* stated, "There is uniform agreement on the neurologic examination with exception of the apnea test. However, this survey found other major differences in the procedures for diagnosing brain death in adults. Standardization should be considered" (Wijdicks 2002).

Uniform Determination of Death Act (UDDA) of 1981,[7] which defines death as either the "irreversible cessation of circulatory and respiratory functions," or the "irreversible cessation of all functions of the entire brain."[8]

We may wonder why the brain came to assume such importance in the diagnosis of death. The conventional medical rationale is that the brain coordinates all vital bodily functions, such that when it dies a total system collapse takes place. While the body may limp along for a short time thereafter with mechanical support, it is believed to have irreversibly lost the capacity for integrated functioning, which is the hallmark of a living organism.

With the passage of time, however, the thesis that brain death equals somatic death has become less convincing, and research has shown that the brain does not, as commonly thought, coordinate all vital bodily functions. For example, while the brain stem helps regulate heartbeat, it does not cause it: the heart has its own internal pacemaker and can continue beating for some time even when removed from the body. In the case of the lungs, the ventilator simply introduces oxygenated air, while respiration (the exchange of gases with the environment) continues at the cellular level independently of the brain.[9]

Evidence of this kind suggests that the loss of function in the brain is not equivalent to the biological death of the body. Significant numbers of brain dead patients have survived for weeks, months, and even years—in one case for 20 years (Repertinger et al. 2006)—and perhaps the numbers would be even greater if a diagnosis of brain death were not the self-fulfilling prophesy it currently is. Resources are rarely expended in the care of such patients,[10] but when they are, the mean

[7] Section 1 of the Act entitled "Determination of Death" states: "An individual who has sustained either (1) irreversible cessation of circulatory and respiratory functions, or (2) irreversible cessation of all functions of the entire brain, including the brain stem, is dead. A determination of death must be made in accordance with accepted medical standards."

[8] While the definition of death in the UDDA seems clear and straightforward, critics have pointed to both conceptual and practical problems. *Irreversibility*, for example, is a prognosis, not a demonstrable fact, and the requirement for the *cessation of all functions* is seen by many as unduly strict because clusters of brain cells often show sporadic activity which may be little more than mental static (Lock 2002, 110ff). Nor is the diagnosis of brain death a simple matter, as a glance at the American Academy of Neurology Guidelines for Brain Death Determination will reveal. http://surgery.med.miami.edu/laora/clinical-operations/brain-death-diagnosis. Accessed 17 December 2016. For a comprehensive review of the various arguments around brain death see *The President's Council on Bioethics, Controversies in the Determination of Death* (2008). https://bioethicsarchive.georgetown.edu/pcbe/reports/death/. Accessed 24 December 2016.

[9] The neural regulation of body temperature also continues, and the spinal cord and peripheral nervous system still function. Essential neurological functions also continue in the brain itself, such as the regulated secretion of hypothalamic hormones. EEG activity is detected in around 20% of brain-dead patients, and when an incision is made to retrieve organs the patient displays a cardiovascular response to stress in the form of increased blood pressure.

[10] Research by Dr. Alan Shewmon contradicts the orthodox view that a brain-dead patient can survive for a few days at the most. Shewmon investigated 175 cases of which 56 showed that brain dead patients can survive considerably longer. Half survived more than a month, a third more than 2 months, 13% more than 6 months, and 7% more than a year. One exceptional case survived for over 14 years (Meyer 2005, p. 16).

survival rate can be extended from days to weeks, as Japanese researchers have demonstrated (Lock 2002, p. 145).[11] Advances in this direction may continue, and in 2016 the US clinical trials authority gave approval for the test of a protocol for reversing brain death. The CEO of the American biologics company undertaking the research (Bioquark) stated, "It is a long term vision of ours that a full recovery in such patients is a possibility" (Cook 2016). This may be overoptimistic, but if brain death is shown to be reversible it will undermine the current justification for the transplantation of organs from brain dead donors.

The urgency with which organs need to be harvested, moreover, means that the standard protocols are not always followed.[12] Misdiagnosis of brain death, or "false positives" as they are known, are not uncommon, and there have been many cases of patients coming "back from the dead" (Schaller and Kessler 2006.). In one such case in 2009, patient Colleen Burns woke up on the operating table of St Joseph's Hospital Health Center in Syracuse, New York just as doctors were about to remove her organs.[13] It transpired that following a drug overdose the patient had lapsed into a coma which doctors had mistakenly diagnosed as brain death. The patient was subsequently discharged from hospital 2 weeks later, with her organs intact.

While the concept of brain death continues to be robustly defended by medical bodies, many both inside and outside the profession feel less confident than before about the claim that the loss of function in the brain is equivalent to bodily death. There is a growing body of dissident literature in medical journals and elsewhere that suggests that the criterion of brain death is conceptually and scientifically flawed,[14] and even some leading supporters of transplantation have accepted it is no longer coherent.

Writing in the *New England Journal of Medicine*, pediatric anesthetist Dr. Robert D. Truog has suggested that "the medical profession has been gerrymandering the definition of death to carefully conform with conditions that are most favorable for transplantation" (Truog and Miller 2008, p. 675). He notes:

> After all, when the injury is entirely intercranial, these patients look very much alive: they are warm and pink; they digest and metabolize food, excrete waste, undergo sexual maturation, and can even reproduce. To a casual observer, they look just like patients who are receiving long-term artificial ventilation and are asleep. (Truog and Miller 2008, p. 674)

[11] Dr. Hayashi Narayuki pioneered the "Hayashi hypothermia technique" and believes that future advances will help prevent brain cells dying after major trauma (Lock 2002, p. 277).

[12] For a study of pediatric donors, see Verheijde et al. (2008) and Rady et al. (2008).

[13] http://abcnews.go.com/Health/patient-wakes-doctors-remove-organs/story?id=19609438. Accessed 17 December 2016. This is not the only such case, see also http://www.telegraph.co.uk/news/newstopics/howaboutthat/2106809/Dead-man-wakes-as-transplant-surgeons-prepare-to-remove-his-organs.html. Accessed 17 December 2016. Lock mentions others (2002, p. 76ff, 54–6, 111, 363). It is difficult to be certain whether such "false positives" are caused by human error in following the protocols—the explanation favoured by the medical profession—or because the assumptions about death on which the protocols rest are themselves flawed. In either case, the consequences for the patient are the same.

[14] E.g., Potts et al. (2001).

"The arguments about why these patients should be considered dead," Truog adds, "have never been fully convincing." In a similar vein, an editorial in *Nature* in October 2009, openly recognizing the ambiguity surrounding current definitions of brain death, stated: "The time has come for a serious discussion on redrafting laws that push doctors towards a form of deceit." "Ideally", it added, "the law should be changed to describe more accurately and honestly the way that death is determined in clinical practice" ("Delimiting Death").

As we saw at the start, many intensivists already believe that brain death is merely the harbinger of bodily death. They locate the significance of brain death rather in the permanent loss of consciousness that accompanies it. Thus, although the patient's body remains alive, the "person" who was the patient no longer exists (Lock 2002, p. 249).[15] The UDDA, however, makes no mention of consciousness or the notion of a "person" who dies separately from their body. Some bioethicists and physicians think it should, and that the law should adopt a new definition of death as "cognitive," "upper brain," or "neo-cortical" death. On this basis death would be defined as an event that takes place when consciousness is permanently lost, thus abandoning any suggestion that death is biological in nature. This would allow a permanently unconscious patient to be declared dead,[16] even though the lower brain may be functioning, the heart beating, and respiration continuing either with or without mechanical support.[17] It would also mean, somewhat counterintuitively, that human death was different from the death of other animals, such as cats and dogs.

Underlying the argument for cognitive death is a dualistic view of human nature that sees "persons" as distinct from their bodies. Philosophers who reject this view, such as Hans Jonas, believe that even if the higher functions of personhood are seated in the brain, "My identity is the identity of the whole organism" (quoted in Lock 2002, p. 95). On this understanding, the body is an integral and unique part of a person's identity, as evidenced by fingerprints and DNA. One dissident intensive care specialist summed this up by saying that his person was "as much embodied in his size nine feet as in his brain" (Lock 2002, p. 249).

The proposal that human death should be redefined as the irreversible loss of consciousness raises philosophical questions as much as medical ones. Where does

[15] Many doctors seem unclear about the criteria for brain death. Research by Stuart Youngner and colleagues published in 1989, 8 years after the UDDA became law, showed that only 35% of respondents correctly identified the legal and medical criteria for brain death, and over half (58%) did not consistently use a coherent concept of death (Lock 2002, p. 123). Lock suggests that the ambiguities and contradictions identified in this 1989 research were still present in 2002 (2002, p. 248).

[16] A leading proponent of this view is Veatch (1976, 2008, 2015). Writing in the *New England Journal of Medicine* (2008), Veatch claims that perhaps a third of Americans support a higher-brain or consciousness-based definition of death. He suggests an amendment is needed to the "dead donor rule" (the principle that organs should only be removed from a dead donor) to allow transplants from patients who are still alive but permanently unconscious.

[17] See, for example, the views of Truog and Miller (2008) mentioned above.

Buddhism stand on the matter?[18] I will suggest Buddhism does not support the notion of cognitive death because in common with most religions it rejects mind-body dualism, and regards mind and body as two aspects of a single reality—like a mixture of milk and water, as Tibetan sources express it. In the following section we explore the background to this belief.[19]

11.3 Classical Buddhist Teachings on Death[20]

Buddhist monks took an interest in medicine from the earliest times and contributed much to the development of the indigenous Indian system of medicine known as Āyurveda (Zysk 1991, p. 6). Monks would have been especially familiar with the stages of death and decomposition because of an ancient practice known as the "cemetery meditations."[21] These meditations took as their object corpses in various stages of decomposition. The purpose of this exercise was not to study anatomy—in fact anatomy never developed as a branch of Asian medicine—but to reinforce awareness of the brevity of life and decrease attachment to the body. It seems likely, however, the practice would also have informed the early Buddhist understanding of death as the beginning of a gradual and irreversible process of bodily decomposition.

The onset of this process, according to classical Indian Buddhist sources, is marked by the disappearance of three factors from a living body. The three are vitality (*āyu*), heat (*usmā*), and consciousness (*viññāṇa*) (S.iii.143). By "consciousness" here is meant not cognitive awareness but a deeper and more diffuse form of organic consciousness that is the basis of sentiency in all its modes. From a Buddhist perspective, there is no single seat of consciousness, whether in the brain or anywhere else (Sugunasiri 1995). Instead, consciousness is thought to suffuse the body in the

---

[18] Meyer (2005) suggests Buddhist sources support the concept of "cognitive death." The textual evidence for this, however, is weak. Essentially it relies on a claim that *cetanā* (intention) can be understood as "synonymous with life itself" (p. 11). To equate *cetanā* with "life itself," however, goes far beyond the normal meaning of the term. More likely, *cetanā* in this context is a cognate of *citta*, which in turn is a synonym for *viññāna* (e.g., *Satipaṭṭhāna Sutta*, S.ii.95). For further discussion, see Sugunasiri (1995). The rest of Meyer's argument relies on the claim that since only humans can attain nirvana, a separate definition of death (a cerebral one) is required in their case. This seems a *non-sequitur*. While Buddhism certainly recognizes the distinctiveness of a human rebirth, it does not follow that unique criteria are required to determine human death.

[19] For an argument that alleged East-West cultural differences in reality play only a minor role in bioethics see Beauchamp (2015).

[20] I will employ the following abbreviations below in my references to the Pali Canon: D = *Dīgha Nikāya* (Rhys Davids and Carpenter 1889); M = *Majjhima Nikāya* (Trenckner and Chalmers 1888); S = *Saṃyutta Nikāya* (Feer and Davids 1884); SA = *Sāratthappakāsinī* (Woodward 1929). See the reference section for full bibliographic information.

[21] See, for example, the *Satipaṭṭhāna Sutta*, http://www.accesstoinsight.org/lib/authors/soma/wayof.html#discourse, and the *Maraṇassati Sutta*, http://www.accesstoinsight.org/tipitaka/an/an06/an06.020.than.html. Both accessed 17 December 2016.

way that moisture suffuses a sponge. Mental awareness is classified as one of six fields of awareness, the other five corresponding to the conventional five senses. The loss of mental awareness is therefore the loss of one field of awareness rather than the loss of the human person. For this reason, it would be a mistake from a Buddhist perspective to take the absence of cognitive awareness as evidence of death (Sugunasiri 1990).

What about the other two factors—vitality and heat? In modern terms, vitality *(āyu)* seems to correspond to the metabolic processes that take place in the body, and heat (*usmā*) to the energy these processes liberate. An obvious way to test when bodily metabolism has ceased, therefore, is by monitoring bodily temperature.[22] Bodily cooling is strong evidence of death, but is there no earlier confirmation we might use, like the absence of heartbeat and respiration?

I think the reason heartbeat and respiration are not mentioned by the early sources has to do with Buddhist meditational practice, and in particular the knowledge that individuals could enter trance-like states resembling death and remain there for a considerable length of time with no sign of either pulse or breathing. Examples include the elder Mahanāga who reportedly remained seated in trance while the meditation hall burnt down around him.[23] This profound state of trance was known as the "state of cessation" (*saññāvedayitanirodha*), and the phenomenon of individuals entering this state is what provoked reflection on how to distinguish between a person who is alive and one who is dead (*mato kālakato*). The conclusion was that this distinction could be made by reference to the three factors mentioned.[24]

The death of the founder of Buddhism provides a further interesting example. The sources report that as the Buddha lay dying he ascended through eight levels of trance *(jhānas)* and attained this state of cessation. At this point his personal attendant, Ānanda, declared that the Buddha had passed away because, as the commentators explain (SA.i.223), he saw no sign of breathing. He was corrected, however, by a senior monk, the Venerable Anuruddha, who informed him that his master had not yet passed away but had merely attained the state of cessation.[25] The existence of this phenomenon—a state in which the subject is alive but where the body generates no vital signs—presents an obstacle to any methodology that claims to define the moment of death with precision. This explains the reluctance on the part of the early sources to accept anything other than the loss of bodily heat as confirmation of death.

[22] Bodily cooling is a widely recognized concomitant of death and is known as *algor mortis*, the process by which the temperature of a body drops from its normal 37° centigrade, assuming normal conditions, until it reaches the ambient environmental temperature. Further observable signs include skin pallor, changes in the eyes such as loss of pressure and marking of red blood cells, flaccidity in the primary muscles, lividity or *livor mortis* (the process of blood flowing downwards and causing a reddish-purple colour on the skin), *rigor mortis* which sets in 3–4 h after death and lasts between 36 and 48 h. Someone observing these phenomena progressively would have little doubt that death had taken place.

[23] The story is reported by the fifth-century commentator Buddhaghosa at *Visuddhimagga* 706.

[24] The *Mahāvedalla-sutta* of the Majjhima Nikāya, M.i.296.

[25] D.ii.156.

Before leaving the early teachings, a final point that bears mention is that for all who have not attained nirvana, death is believed to be the gateway to a new rebirth, and the circumstances of death are thought to have an important bearing on the condition of rebirth in the next life. A peaceful conscious death is generally seen as the best way to die, and a confused, painful, or traumatic death—such as while having one's vital organs removed—as one of the worst. The intensive medical intervention required for transplantation means that organ donors will not die peacefully with relatives at their bedside, which is the kind of death most donors perhaps imagine they will undergo before their organs are harvested.

11.4 Contemporary Attitudes

11.4.1 Japan

We turn now to a consideration of contemporary attitudes in the two countries mentioned at the start, beginning with Japan. Understanding of the Japanese view of brain death and organ transplantation was advanced by a study published in 2002 by anthropologist Margaret Lock. The title of this award-winning work was *Twice Dead: Organ Transplants and the Reinvention of Death.* Near the start of the book Lock describes the issue of brain death (*nōshi mondai*) in Japan as "the most contentious ethical debate of the last thirty years" (2002, p. 3). In contrast to the West, brain death is regularly discussed in the Japanese media, in popular books, and even features in manga comics.

Transplant medicine in Japan initially suffered a setback in what became known as the "Wada case." This concerned doctor Wada Juro who in 1968 carried out the world's thirtieth heart transplant. The recipient died 83 days after the operation, and the ensuing investigations revealed no evidence to support the diagnosis of brain death, or even that a heart transplant was required. It transpired that Wada had lied about various details of the procedure and evidence had been tampered with. This resulted in charges of homicide and professional negligence that were eventually dropped, but the furor rumbled on. Such was the public outcry that it became impossible to perform further transplants from brain dead patients, and brain death was not legally recognized for almost three decades thereafter.

The Japan Medical Association eventually approved the concept of brain death in 1988, but it was not until the 1997 Organ Transplant Law (amended in 2009) that the legal validity of brain death was recognized. According to Yasuoka, however, "There is still no formal definition of brain death in Japan" (2015, p. 15). Official recognition of brain death, moreover, is limited to the specific context of cadaver organ transplantation, and in contrast to the UDDA the Japanese law does not recognize brain death as a universal standard of death. The rate of organ transplants in Japan is the lowest in the industrialized world, with less than a hundred operations performed in 2013.[26] A consequence of this is that, as Yasuoka reports, "Japan has the severest organ shortage in the world" (2015, p. 6), a fact that leads many Japanese to travel abroad for operations.

Lock notes how a poll carried out in 1987 showed that only 24 percent of the public thought that brain death meant the end of life (2002, p. 137), a view in keeping with the traditional Japanese belief that death does not take place until the soul leaves the body. This is thought to occur "when the body becomes cold and starts to stiffen" (Lock 2002, p. 198), a view similar to the one we encountered in classical Indian sources.

The traditional Japanese concept of the "person," furthermore, is as co-extensive with the body rather than merely with the mind and brain. In East Asian medicine the idea of a life force—*qi*—distributed throughout the body rather than located in any single organ is deeply ingrained, and underlies the practice of acupuncture and martial arts (Lock 2002, p. 199). Yonemoto Shohei, a well-known Japanese commentator, notes how in contrast to "Americans who think of organs as replaceable parts, … Japanese tend to find in every part of a deceased person's body a fragment of that person's mind and spirit" (Lock 2002, p. 226). Most Japanese believe that their true inner self (*kokoro*) lies in the depth of the body and is not confined to any one organ. "Such views," writes Lock, "make it difficult to count brain-dead persons as dead, particularly when a brain-dead body remains so visually alive" (2002, p. 228).[27]

Buddhists make up about one third of the population of Japan, and Buddhism is regarded as having special expertise in matters of death and dying. There is a Japanese saying that "one is born in Shinto and dies in Buddhism." The tenth-century Buddhist priest Genshin composed a list of "Rules for Dying" (*Rinjū no gyōgi*), and a series of fourteenth-century pictures (the *Kyusōshi emaki*) depicts nine stages of dying. In the last two, maggots eat the body leaving only the bones (Lock 2002, p. 296), in an echo of the ancient Indian "cemetery meditations."

---

[26] According to information published in the *Japan Times* (2014), "In the United States, where organ transplantation is better accepted, there are 7000–8000 organ transplants every year, which works out to about 26 organ transplants per million population. Contrast that to Japan, where the rate is just 0.9 transplants per million, the lowest rate in the industrialized world. Fewer than 100 organ transplants (*zōki ishoku*) were performed in Japan last year." "Organ Donation" July 18, 2014. http://www.japantimes.co.jp/news/2014/07/18/reference/organ-donation/#.WEhbmL0rLIU. Accessed 27 December 2016. Writing in 2006, Ronald Y. Nakasone reports "approximately 7000 persons are determined to be brain dead every year and 13,072 persons are on waiting lists to receive organs" (2006, p. 291). The number of operations quoted varies, but figures overall seem to be in the low hundreds. The English portion of the website of the Japan Organ Transplant Network gives only limited information and few statistics. http://www.jotnw.or.jp/english/index.html. Accessed 27 December 2016. Yasuoka reports that as of October 30, 2015 "only 290 operations have been completed" (2015, p. 1).

[27] The writings of Japanese philosopher Morioka (1989) are interesting in this respect. A selection of his writings on bioethics (in English) can be found online at www.lifestudies.org. Accessed 17 December 2016.

Until recently, however, the voice of Buddhism has not been much heard on the specific question of brain death. In 1990 a report from the Japanese Association of Indian and Buddhist Studies concluded simply with an appeal to the medical profession to reach a consensus.[28] In 1994 Helen Hardacre, a Harvard Japanologist, noted that "[c]ompared with the volume and variety of debate elsewhere [in Japan], the response from Buddhism and Shinto has been almost negligible" (Lock 2002, p. 210). Since then things have moved on. In 2005 a research panel on bioethics was established by an umbrella organization of major Japanese Buddhist groups known as the Forum of Research Institutes Associated with Religious Organizations (*Kyōdan Fuchi Kenkyūsho Konwakai*). The issues of brain death and organ transplantation were chosen as themes for its inaugural meeting, reflecting public concern around these topics. Ugo Dessì, a Japanologist from the University of Leipzig, summarized the discussion as follows:

> As it may easily be observed, the various viewpoints presented by the participants to the meetings of the Forum have generally insisted on the problematical nature of the notion of brain death, and from the beginning a general consensus has been achieved on the fact that this stage cannot be uniformly accepted as the death of the individual ... Most research institutes have insisted that the true indicators of the death of a human being are instead the three traditional signs ... namely, cardiac arrest, respiratory arrest, and dilation of the pupils. (Dessì 2010, p. 95)

Dessì adds that this understanding of the nature of death involves a rejection of "the dualism between body and soul typically expressed by the western philosophical tradition" (2010, p. 95) and concludes "almost all the participants agree that brain death cannot be accepted as the actual death of the individual without exceptions" (2010, p. 98). Other religions in Japan seem to concur, and in a 2006 Symposium on Religion and Bioethics held in Tokyo, the Japanese Association of Religious Organizations reportedly achieved "complete consensus" on the principle that "brain death may not be recognized as human death in all individual cases" (Dessì 2010, p. 93).[29]

[28] Japanese Association of Indian and Buddhist Studies Committee for Enquiry on Brain Death and Organ Transplantation, September 1990. Reprinted in Fuji (1991). Other sources from the early 1990s are cited by Veatch and Ross (2015, p. 20 n30).

[29] There appear to be only two dissenting voices to this consensus. One is the agnostic view of a research institute representing Zen Buddhism which holds that "Zen Buddhism does not provide a definitive argument for the acceptance or refusal of organ transplant" (Dessì 2010, p. 94). The other is Soka Gakkai, Japan's largest lay Buddhist organization. A 1994 article on their website by immunologist Dr. Yoichi Kawada states, "Current medical technology is incapable of reviving people who have reached the stage of 'brain death,' and my understanding of Buddhism can accept this as the present-day meaning of death" (http://www.sgiquarterly.org/feature2004Apr-2.html. Accessed 15 December 2016). The author has published a book on the subject in Japanese (Kawada 1996). It is reported that this group has also argued in favour of donor cards and the establishment of an information network for organ donors (Lock 2002, p. 210n). I have been unable to obtain access to an early statement by Ikeda (1987) which may provide more details on Soka Gakkai's position.

11.4.2 Thailand

We turn now to Thailand, which is similar to Japan in two respects and different in another. The similarities are the low number of cadaver transplants, and a loss of public confidence due to medical scandals. The difference is that whereas the legitimacy of the concept of brain death has been much debated in Japan, in Thailand it has been accepted with little public discussion. According to one researcher, "in Buddhist Thailand there has apparently been no resistance either to the recognition of brain death or to the donation of human organs" (Ohkubo, quoted in Lock 2002, p. 329).

Organ transplantation from both living and dead donors is carried out in Thailand, which has a 95% Buddhist population. The first organ transplant was a kidney transplant carried out at the King Chulalongkorn Hospital, Bangkok in 1972. In due course criteria for brain death were established by a committee of physicians in 1989 (revised in 1996). These were given legal status in the Observance on Medical Ethics 1995 (and subsequent amendments), an ordinance promulgated under the Medical Professional Act of 1992.

The Organ Donation Center is a charity established in 1994 to register hospitals, manage waiting lists, register prospective donors, and serve as a forum for collaboration and the exchange of information.[30] In 2017, its director, Dr. Visist Dhitavat, reported that his center registered just 220 organ donors in the previous year and their organs were allocated to 512 patients. The total number of patients waiting for organ donations in the same year was 5581 (Saengpassa 2017). Dhitavat is earlier reported as saying that "most Thais did not want to donate their organs because they believed those body parts would be missing if they were reincarnated" (Fernquest 2011). In the same newspaper report Sophon Jirasiritham, chairman of Ramathibodi Hospital's kidney transplant project, located the problem elsewhere and said, "better management of trauma cases was needed if the brain-death stipulation was to be applied more often." He added, "State hospitals and medical schools should find ways to enable their doctors and nurses to better determine brain death, as this could help patients in need of donated organs."

What might be called "transplant tourism" is part of the broader phenomenon of "medical tourism," and Thailand is a leading provider of medical services to foreigners, reportedly earning some 850 million dollars in 2008.[31] Rumors about the sale of organs in Thailand and other Asian countries began to circulate in 1989, and were confirmed in 1999 when a private transplant hospital, Vajiraprakan Hospital in Samut Prakan province, was found to have violated a number of legal requirements including the purchasing of kidneys. An investigation by the Thai Medical Council revealed violations of the rules for brain death certification, as in some cases only

[30] The information on the Thai Transplantation Society website seem out of date, but speaks of 400 kidney transplants a year. Historical data suggests perhaps half of these are from brain-dead donors. http://www.transplantthai.org/en/transplant24-00005.jsp. Accessed 17 December 2016.

[31] Teh (2007, p. 191) quotes a figure of US$615 million in 2005.

one physician had signed the certification and the same doctor was involved in the transplant. Payments had been made to relatives of the deceased, and there was no evidence of kinship for nine out of 35 supposedly related donors. "Kick-back" payments were made to ambulance services for transferring accident victims, as well as to neighboring hospitals for transferring potential brain dead patients. According to Tungsiripat, this had "major repercussions on public trust in transplantation" and the number of donations and transplantations decreased significantly (2003, p. 8). A number of those involved were subject to professional sanctions, and charges of murder and forgery were brought (Teerawatanon et al. 2003). After lengthy legal proceedings the accused, three doctors and a former manager of the hospital, were acquitted in September 2016.[32]

The issue that has attracted most comment in Thailand, however, is not organ transplantation but the withdrawal of life support. This became a point of public controversy following the death of the renowned monk and teacher Buddhadasa in 1993 following a stroke. Debate raged over whether the monk should have been allowed to die peacefully in his forest monastery, as he had specified in an advance directive, or whether efforts should have been made to prolong his life (Ratanakul 2000; Jackson 2003; Kanjanaphitsarn 2013, 2015). The case highlighted the reluctance of Thai physicians to withdraw life support from terminal patients despite the legal acceptance of brain death,[33] and the existence of a patients' "bill of rights" (The Thai Medical Council 2000) authorizing patients to make decisions about their medical care.

In practice, Thai physicians will generally not remove breathing tubes from patients themselves, although some will allow a family member to do it. Stonington and Ratanakul explain that Thai physicians "have a complex array of reasons for declining to remove ventilator support, including their medical training, fear of litigation, and belief in the sanctity of life" (2006, p. 1680). There is also a common Thai belief that the last part of the body to die is the breath, and hence "pulling out a patient's ventilator may feel like pulling out the patient's soul" (2006, p. 1680). A physician who performs such an action, it is thought, inevitably incurs spiritual demerit.

This reluctance to withdraw life-support has led to an accumulation of patients being kept alive by machines. Stonington describes one ward where "half the patients were on mechanical ventilators. The air was sterile and filled with the beeps of machines. Nurses scuffled around with gloves, wheeling blood-pressure check units to the beds of almost corpse-like patients, strapped as modern cyborgs into the life-machines of medical innovation" (2012, p. 841). In view of the strain it places on resources, this is not a sustainable position. As Stonington and Ratanakul note,

[32] http://www.bangkokpost.com/archive/doctors-acquitted-of-murder-for-organ-transplants/1097096. Accessed 27 December 2016.

[33] Thai philosopher Hongladarom (n.d.) suggests Buddhism would go along with whatever definition of death a society chose to adopt provided it was in accordance with Buddhist tenets, but leaves unexamined the question of whether brain death is, or is not, in accordance with Buddhist tenets.

"there is an urgent need for solutions to the "ventilator problem"—both to patch the failing universal healthcare system and to help Thais make difficult decisions about intervention at the end-of-life" (2006, p. 1681).

As noted in Sect. 11.2, the first of the two stated purposes of the brain death criterion was to allow the discontinuation of life-prolonging treatment. In the West, a diagnosis of brain death is often invoked as a justification for the discontinuation of treatment in circumstances of the kind described, and it is curious that Thai doctors do not make similar use of it.[34] This may be due to a lack of training, as suggested by Dr. Jirasiritham above, although this does not explain why such training has not been provided if the problem is so urgent. An alternative explanation, which may also account for the low levels of organ donation, is that the apparently unproblematic acceptance of brain death at an official level disguises a deeper, and as yet little explored, incompatibility with indigenous beliefs. Further research would be required to confirm this hypothesis.

We can ask, meanwhile, whether there is a Buddhist solution to the Thai "ventilator problem." This is a question I have addressed more fully elsewhere in the context of Euthanasia (2018), but an initial response might be to suggest that a patient who can only be kept alive with the assistance of a ventilator is in fact already dead on the traditional cardiopulmonary criterion because their breathing is artificially maintained. A problem with this view is that patients with cervical paraplegia also require ventilator support, and patients with chronic renal failure are also kept alive by machines. The mere fact of dependency on artificial life-support, therefore, is not in itself a useful criterion for determining when treatment should be withdrawn.

A more successful line of argument might be that the treatment being administered (artificial ventilation) is disproportionate, in other words either futile or too burdensome in light of the patient's condition and prognosis. It is accepted in Western medical practice that it is pointless to continue a treatment that cannot ameliorate the patient's condition, and that a time may come when the patient is simply beyond medical help given the limitations of current medical knowledge and the resources available. Determining when this point has been reached requires experience and judgment, and there is no mathematical formula that can be used to define it with precision. It is important, however, that both doctors and patients recognize when this point has been reached so that medical treatment does not become an intrusive and futile struggle against death itself. This is particularly so where resources are scarce, and there are others who might be helped more effectively. The solution to the Thai problem may accordingly lie in recognizing the concept of futility—the acceptance that nothing more can be done therapeutically—as a justification for the withdrawal of life support. A diagnosis of brain death would then be one piece of evidence to be weighed in the balance, but ventilator support would be withdrawn on the grounds that it was no longer of therapeutic value to the patient rather than because the patient was deemed to be dead. This is not to say that such support *must* be withdrawn, only that it would not be morally obligatory to continue it.

---

[34] This justification would be unavailable in Japan where the concept of brain death is only legally valid in the context of cadaver organ transplantation.

11.5 Conclusion

Above we explored the discrepancy between the modern concept of brain death and the traditional Buddhist understanding of death as the loss of the body's organic integrity. According to the classical Indian sources, there is no "magic moment" at which life ends. This casts doubt, at least in my construction of the Buddhist view as described above, on the claim that death can be diagnosed with precision by reference to the cessation of function of a single organ. The further claim that death should be redefined as "cognitive death," or the permanent loss of consciousness, was also rejected as incompatible with Buddhist teachings on the non-dualistic relationship between mind and body. In the two countries discussed we noted similarities and differences: in Japan brain death was given statutory recognition after intense debate, whereas in Thailand its legal recognition occurred with little public discussion. In neither case, however, does the concept appear to have been well received at a popular level.

If the analysis provided thus far is accurate, Buddhists face a difficult choice. A refusal to participate in cadaver organ transplantation programs means that lives will be lost, since there are few, if any, alternative treatments for major organ failure. The development of artificial organs, and the use of stem cells supplied by the patient to grow replacement organs, are promising ways forward, but still some way off. Xenotransplantation (transplanting organs from animals) and the growing of human organs inside animal bodies are further possibilities.[35] None of these techniques will deliver results in the short term, and so as long as transplantation is seen as the primary solution to organ failure the alternatives are unlikely to receive the attention and funding they deserve.

A final question concerns public trust in the medical profession. Brain death has become something of a medical dogma, an article of faith for the profession, and in spite of its defects has become widely accepted in the West due to the public's trust in health care professionals. This trust can easily be dented, as has happened in Japan and Thailand, and it will not serve the interests of either doctors or patients to sustain a flawed criterion of death in the longer term.

References

Beauchamp, Tom. L. 2015. Common morality, human rights, and multiculturalism in Japanese and American bioethics. *Journal of Practical Ethics* 3: 18–35.

Cook, Michael. 2016. Bold attempt to reverse brain death gets US approval. *Bioedge*. May 7. https://www.bioedge.org/bioethics/bold-attempt-to-reverse-brain-death-gets-us-approval/11862. Accessed 17 Dec 2016.

Delimiting death. 2009. *Nature* 461: 570.

[35] As reported at http://www.bbc.co.uk/news/health-36437428. Accessed 31 December 2016.

Dessì, Ugo. 2010. Religion, networking, and social issues in Japan: The case of the Kyōdan Fuchi Kenkyūsho Konwakai. *Japanese Religions* 35 (1–2): 87–100.

Euthanasia. 2018. In *The Oxford handbook of Buddhist ethics*, ed. Daniel Cozort and James Mark Shields. Oxford: Oxford University Press.

Feer, L., and C.A.R. Davids. 1884. *The Saṃyutta- Nikāya of the Sutta-Piṭaka*. Vol. 6. Oxford: Pali Text Society.

Fernquest, John. 2011. Organ donors save lives, *Bangkok Post*, August 15.

Fuji, M. 1991. Buddhist bioethics and organ transplantation. *Okurayama Bunkakaigi Kenkyunenop* 3: 1–11.

Hongladarom, Soraj. n.d. *Organ transplantation and death criteria: Theravada Buddhist perspective and Thai cultural attitude*. http://pioneer.chula.ac.th/~hsoraj/web/Organ%20 Transplantation-Buddh.pdf. Accessed 18 May 2017.

Ikeda, D. 1987. Thoughts on the problem of brain death (1): From the viewpoint of the Buddhism of Nichiren Daishonin. *Journal of Oriental Studies* 2: 193–216.

Jackson, P.A. 2003. *Buddhadasa: Theravada Buddhism and modernist reform in Thailand*. Chiang Mai: Silkworm Books.

Kanjanaphitsarn, S. 2013. An analytical study of euthanasia in Buddhism with special reference to the case of Buddhadāsa Bhikkhu's death. *International Journal of Buddhist Thought & Culture* 21: 141–154.

———. 2015. *Euthanasia in Buddhism: A case of Buddhadasa Bhikkhu's death*. Saarbrücken: Scholars' Press.

Kawada, Yoichi. 1996. *Noshi mondai to Bukkyo shiso* [The problem of brain death and Buddhism]. Tokyo: Daisan Bunmei-sha.

Keown, D. 2001. *Buddhism and bioethics*. London: Palgrave.

———. 2010. Buddhism, brain death and organ transplantation. *Journal of Buddhist Ethics* 17: 1–36.

Lock, M. 2002. *Twice dead: Organ transplants and the reinvention of death*. Berkeley: University of California Press.

Meyer, John-Anderson. 2005. Buddhism and death: The brain-centered criteria. *Journal of Buddhist Ethics* 12: 1–24.

Morioka, Masahiro. 1989. *Brain dead person: Human relationship oriented analysis of brain death*. Lifestudies.org. http://www.lifestudies.org/braindeadperson00.html. Accessed 27 Dec 2016.

Nakasone, Ronald Y. 2006. Ethics of ambiguity: A Buddhist reflection on the Japanese transplant law. In *Handbook of bioethics and religion*, ed. David E. Guinn, 291–303. Oxford University Press: Oxford.

Potts, M., P.A. Byrne, and R.G. Nilges. 2001. *Beyond brain death: The case against brain based criteria for human death*. Berlin: Springer.

Rady, M.Y., J.L. Verheijde, and J. McGregor. 2008. Organ procurement after cardiocirculatory death: A critical analysis. *Journal of Intensive Care Medicine* 23 (5): 303–312.

Ratanakul, P. 2000. To save or let go: Thai Buddhist perspectives on euthanasia. In *Contemporary Buddhist ethics*, ed. D. Keown, 169–182. London: Curzon Press.

Repertinger, S., W.P. Fitzgibbons, M.F. Omojola, and R.A. Brumback. 2006. Long survival following bacterial meningitis-associated brain destruction. *Journal of Child Neurology* 21: 591–595.

Rhys Davids, T.W., and J.E. Carpenter. 1889. *The Dīgha Nikāya*. Vol. 3. Oxford: Pali Text Society.

Saengpassa, Chularat. 2017. Patients waiting for organ transplants face shortages. *The Nation*, April 3. http://www.nationmultimedia.com/news/national/30311097. Accessed 18 May 2017.

Satipaṭṭhāna Sutta. n.d.. http://www.accesstoinsight.org/lib/authors/soma/wayof.html#discourse.

Schaller, C., and M. Kessler. 2006. On the difficulty of neurosurgical end of life decisions. *Journal of Medical Ethics* 32: 65–69.

Stonington, S.D. 2012. On ethical locations: The good death in Thailand, where ethics sit in places. *Social Science & Medicine* 75: 836–844.

Stonington, S., and P. Ratanakul. 2006. Is there a global bioethics? End-of-life in Thailand and the case for local difference. *PLoS Medicine* 3: 1679–1682.

Sugunasiri, S.H.J. 1990. The Buddhist view concerning the dead body. *Transplantation Proceedings* 22: 947 949.
Sugunasiri, Suwanda. 1995. The whole body, not heart, as "seat of consciousness": The Buddha's view. *Philosophy East and West* 45: 409–430.
Teerawatanon, Yot, et al. 2003. Health sector regulation in Thailand: Recent progress and the future agenda. *Health Policy* 63: 323–338.
Teh, Ivy. 2007. Healthcare tourism in Thailand: Pain ahead? *Asia Pacific Biotech News* 11: 493–497.
The President's Council on Bioethics: Controversies in the determination of death: A white paper by the President's Council on Bioethics. n.d.. https://bioethicsarchive.georgetown.edu/pcbe/reports/death/. Accessed 15 Dec 2016.
The Thai Medical Council. 2000. The declaration of patient's rights. *The Thai Medical Council of Thailand Bulletin* 7: 2–3.
Trenckner, V., and R. Chalmers. 1888. *The Majjhima Nikāya*. Vol. 3. Oxford: Pali Text Society.
Truog, R.D., and F.G. Miller. 2008. The dead donor rule and organ transplantation. *New England Journal of Medicine* 359: 674–675.
Tsomo, Karma Lekshe. 2006. *Into the jaws of Yama, lord of death: Buddhism, bioethics, and death*. Albany: State University of New York Press.
Tungsiripat, Rachata Viroj Tangcharoensathien. 2003. *Regulation of organ transplantation in Thailand: Does it work?* Health economics and financing programme working paper 04/03, LSHTM. https://assets.publishing.service.gov.uk/media/57a08c11ed915d622c0010df/WP04_03.pdf. Accessed 13 Dec 2016.
Veatch, R.M. 1976. *Death, dying, and the biological revolution: Our last quest for responsibility*. New Haven: Yale University Press.
———. 2008. Donating hearts after cardiac death—Reversing the irreversible. *New England Journal of Medicine* 359: 672–673.
Veatch, R.M., and L.F. Ross. 2015. *Transplantation ethics*. Washington, DC: Georgetown University Press.
Verheijde, J.L., M.Y. Rady, and J.L. McGregor. 2008. Growing concerns about brain death and organ donation. *Indian Pediatrics* 45: 883–888.
Wijdicks, E.F.M. 2002. Brain death worldwide: Accepted fact but no global consensus in diagnostic criteria. *Neurology* 58: 20–25.
Woodward, F.L. 1929. *Sāratthappakāsinī*. Vol. 3. Oxford: Pali Text Society.
Yasuoka, M.K. 2015. *Organ donation in Japan: A medical anthropological study*. Lanham: Lexington Books.
Zysk, K.G. 1991. *Asceticism and healing in ancient India: Medicine in the Buddhist monastery*. Oxford: Oxford University Press.

Chapter 12
Ethical Engagement with the Medicalization of Death in the Catholic Tradition

Gerard Magill

Abstract The Catholic tradition can help to guide patients and practitioners through the complex issues that arise due to the medicalization of death because of contemporary medical technology. The purpose is to illustrate how this religious denomination makes moral decisions in practice. The Catholic tradition moors its moral teachings in the constructive interplay between faith and reason, each of which opens itself to the other for insight and enlightenment. The analysis begins with the theoretical realm to discuss the theological foundations and the meaning of Natural Law that guide the *Ethical and Religious Directives for Catholic Health Care Services*. This stance is then applied to several practical concerns that ethically engage issues surrounding the medicalization of death, including maternal-fetal conflicts, patients approaching the end of life, and after death dilemmas. The flexibility and range of Catholic morality can offer astute guidance not only to those who follow this tradition but also to those with different faith perspectives (or none) when they encounter the heart-wrenching dilemmas that medical technology presents around death and dying.

12.1 Introduction

The Catholic tradition offers a distinctive approach to the perennial concerns around dying. This tradition can help to guide patients and practitioners through the complex issues that arise due to the medicalization of death because of contemporary medical technology. This contribution functions both at the theoretical level and the practical level. Often, when faced with ethical dilemmas at the end of life, from early-stage embryos to geriatric patients, families seek practical resolutions that are consistent with the deeper perspectives of religious faith and personal belief. While the Catholic tradition develops its own doctrines, because it merges faith and reason robustly, its stances often can be helpful to many other denominational and secular

G. Magill (✉)
Duquesne University, Pittsburgh, PA, USA
e-mail: magillg@duq.edu

T. D. Knepper et al. (eds.), *Death and Dying*, Comparative Philosophy of Religion 2, https://doi.org/10.1007/978-3-030-19300-3_12

perspectives. From the outset, it can be helpful to recognize that the Catholic tradition moors its moral teachings in the constructive interplay between faith and reason, each of which opens itself to the other for insight and enlightenment. A brief explanation of this general method in Catholicism can help to clarify the practical topics that are addressed subsequently. The explanation considers both the theological foundations and the rational foundations of Catholic morality.

12.2 Theological Foundations

The theological foundations of Catholic morality revolve around an understanding of creation and eschatology—the latter referring to what lies ahead with God after our life on earth. The opening story of the Old Testament depicts the flawed characters of Adam and Eve, with the story of the apple depicting the Fall of humanity from God's grace. As the Old Testament unfolds, its major books project an all-encompassing battle between good and evil whereby humanity determines its fate before God: rejecting God's call in sin, seeking forgiveness from God, being reconciled, and then again falling away to further temptation. Although Christian traditions broadly respect the Old Testament as leading to the New Testament, there is disagreement about the status of humanity because of the original Fall depicted in the Book of Genesis. Generally, Catholicism believes that humanity did not fall totally from God's favor, retaining a core goodness based on being created in God's image, whereas Protestantism typically sees a profound rupture in the Fall that radically separated humanity from God.

Certainly, both traditions join in recognizing that the Son of God in the New Testament brought salvation to humanity, but the bifurcation of belief about the Fall again yields different perspectives. On the one hand, Protestantism perceives salvation as a divine gift that redeems the abject nature of humankind that had totally fallen away from God. On the other hand, Catholicism interprets salvation indeed as redemption, but one that transforms the basic human relation with God that had survived the Fall. These different interpretations present a basic hermeneutic for the way Catholicism and Protestantism address morality. Insofar as Protestantism recognizes little hope for humanity after the Fall until God's redemption in the New Testament, it relies on faith to recognize God's redemptive presence in our lives, celebrating divine grace through personal belief. In contrast, Catholicism retained confidence in human nature after the Fall, as still being in relation with God. Indeed, for Catholicism, the salvation of the New Testament is needed for nature in the following sense: redemption transforms the basic good in human nature to celebrate divine grace in human actions.

This contrast generated a tension between the role of divine faith and human action. This discord led to different interpretations of God's kingdom as either inner worldly (seeking good human action in this life based on reason) or as other worldly (seeking beatitude with God in the next life through faith). This polarized description is provided here to highlight the inherent tension between faith and reason that often characterizes the differences between Protestantism and Catholicism regarding morality. In other words, the good that Catholicism recognizes in humanity after the Fall in the Old Testament, albeit radically transformed through salvation in the New Testament, generated a focus upon human actions based on reason that can lead to God. This focus means using reason to discern what can be good in human action, while aligning that good also with faith to divine salvation.

In sum, this understanding of creation and eschatology in the Catholic tradition provides the theological foundations for its approach to morality: there is inherent good in creation, surviving the Fall in the Old Testament, and being transformed by Redemption in the New Testament. This inherent good in creation can be discerned by reason, albeit aligned with the salvific assistance of divine grace. In Catholicism, reason and faith work together, synergistically, with reason seeking insight into the human good to reveal God's grace in faith. This alignment of faith and reason was argued perhaps most strenuously in the Catholic tradition by John Henry Newman in the nineteenth century (Magill 2015, pp. 73–112).

From the perspective of morality, this basic respect for reason, as working together with faith, is why the Catholic tradition relies on what is known as Natural Law to discern and teach its moral doctrines. The next section explains the significance of Natural Law in the Catholic Tradition (reflecting the theological foundations of creation and eschatology that engender its synergistic view of faith and reason) insofar as it undergirds specific moral teaching around death and dying.

12.3 Natural Law and Double Effect

Natural Law in the Catholic tradition refers to a theoretical approach to morality to guide Church doctrine and practical decision-making. This phrase refers to interpreting nature in a manner that generates moral laws, hence the name of Natural Law. Even after the Fall, humanity's rationality was still inherently good, with a reliable capacity to infer moral duties from God's creation in nature. This capacity was enhanced with the New Testament's revelation of divine salvation. As a result, Catholicism relies not just on biblical revelation to discern morality, but privileges human rationality to infer moral truths from nature. This is a very different approach from mainstream Protestantism, which relies distinctively on the Bible for divinely revealed moral truths.

A closer look at how the Catholic tradition understands and uses Natural Law is indispensable for comprehending its teaching of practical moral dilemmas related to death and dying. There are two main approaches to Natural law in Catholicism, each reinforcing the other. The first approach uses Natural law to infer universal moral truths from human nature. This approach reflects the conservative approach to morality in Catholicism. Both general and specific practical norms are justified from this perspective. On the one hand, general moral truths are elicited from interpreting nature at a theoretical level. For example, the dignity of being human bears a general moral obligation to respect human life. On the other hand, specific moral truths can be elicited from interpreting nature at a practical level. For example, the general obligation to respect human life inspires a pro-life stance to protect the unborn. This practical obligation generates the absolute prohibition of intentionally aborting human life. In other words, this conservative approach to Natural Law leads to a universal moral doctrine that forbids intentional abortion.

The second approach to Natural Law in Catholicism adopts a more historical approach (in contrast to the above universalist approach) that focuses on the individual person (in contrast to the above focus on nature). This second approach presents a more progressive perspective than the conservative stance discussed above. However, this second approach to Natural Law should be construed as being complementary to (not contradictory of) the first approach. An explanation of the details of this progressive approach reveals not only how it is complementary to, but also why it is indispensable for understanding, Natural Law. This second approach focuses upon the distinctiveness of individual persons facing concrete moral dilemmas in very specific historical circumstances. The approach seeks to apply the universal moral doctrines of the conservative approach (such as on abortion) to practical scenarios that require the flexibility of individual discernments about and resolutions of complex moral predicaments.

A closer consideration of the discussion of abortion can be illustrative. On the one hand, often situations arise in pregnancy where the life of the mother and the embryo may be in conflict. A classic case is that of a mother (with a pre-viable fetus) with cancer of the womb. She must have the womb removed for medical treatment if she is to survive, but doing so inevitably leads to the death of the pre-viable embryo. The ethical analysis of the progressive approach to Natural Law argues that the mother as an individual person can be saved if the womb is removed. The justification of doing so is both personal and historical: personal to save the only life that can be saved in the situation, and historical in the sense of facing concrete circumstances that present a dreadful dilemma. The ethical resolution is to understand that foreseeing the inevitable death of the embryo in this terrible conflict is very different from intending the death of the embryo. In other words, the removal of the cancerous womb can be justified as a minimally necessary intervention to treat the mother, even though the inevitable death of the pre-viable embryo is foreseen but not intended.

This case illustrates an approach in the Natural Law tradition that can be aligned with what is known as double-effect reasoning (Magill 2011, pp. 848–867). This approach is articulated in the Principle of Double Effect that involves several conditions: the intervention is either morally good or neutral in itself (the necessary hysterectomy); the bad effect (the death of the fetus) does not cause the good effect (evidenced in recognizing that simply removing the fetus via abortion would not resolve the underlying pathology of the cancerous womb); the good action (the hysterectomy) is what is intended—the death of the pre-viable fetus is unavoidable, foreseen, but not intended; and there is proportion between the good action (the hysterectomy) and the bad side-effect (the death of the fetus) in the sense that the one life that can be saved is saved (to do nothing would lead to the death of both the mother and the fetus).

12.4 Ethical and Religious Directives for Catholic Health Care Services

The theological foundations of the Catholic tradition and the reliance on Natural Law in Catholic morality guide the practical teachings of the U.S. Catholic Bishops on issues related to health care, as articulated in the *Ethical & Religious Directives for Catholic Health Care Services*, published by the U.S. Bishops (hereafter referred to as *ERD*s). These teachings guide all Catholic facilities in the U.S. that provide health care, from hospitals to nursing homes, in accordance with the perspective of Catholic bioethics (Magill 2014, pp. 356–373). The *ERD*s engage clinical issues as well as organizational issues in health care (Magill 2013a, pp. 271–290, 2017, pp. 523–536; Magill and Prybil 2011, pp. 25–50). They are especially important for dealing with issues related to death and dying, which is the focus of this analysis. Before delving into how the *ERD*s offer moral guidance for issues related to death and dying, for which they are perhaps best well known, it can be helpful to explain how they are organized. After a Preamble and a General Introduction that delineate the underlying theology of the document, there are six main sections: on the Social Responsibility of Catholic Health Care Services; on the Pastoral and Spiritual Responsibility of Catholic Health Care; on the Professional-Patient Relationship; on Issues in Care for the Beginning of Life; on Issues in Care for the Seriously Ill; and on Forming New Partnerships with Health Care Organizations and Providers. It is the penultimate section that contributes directly to the discussion of approaches to death and dying in the Catholic tradition.

Each of these main sections of the *ERD*s has a brief Introduction delineating the relevant principles in theology that are applied in the discussion of the practical Directives (referred to below with numbers, such as *ERD*, 56). This relation between the theological narrative in the Introduction and the moral teachings in the Directives is important for understanding the Catholic stance on death and dying. The Introduction of the main section on death and dying in the *ERD*s highlights several related points, reflecting the theological foundations discussed above. Recognizing that God's redemption and salvation "embrace the whole person … in illness, suffering, and death," it is crucial to understand that "we do not have absolute power over life" and "the duty to preserve life is not absolute" (*ERD*, Part Five, Introduction). Hence, it is justifiable to "reject life-prolonging procedures that are insufficiently beneficial or excessively burdensome" (*ERD*, Part Five, Introduction). To honor this moral guidance "about the use of technology to maintain life," two extremes must be avoided: "an insistence on useless or burdensome technology even when a patient may legitimately wish to forgo it and … the withdrawal of technology with the intention of causing death" (*ERD*, Part Five, Introduction).

Before proceeding to the details of Catholic moral teaching on death and dying, it is important to grasp the meaning of a critical remark quoted above that deals with the traditional language of burden and benefit in medicine. The use of the preposition "or" is pivotal in the cited teaching about it being justifiable to "reject life-prolonging procedures that are insufficiently beneficial or excessively burdensome"

(*ERD*, Part Five, Introduction). This teaching adopts a basic distinction that undergirds all Catholic doctrine on death and dying—the distinction between what is called ordinary (or proportionate) and extraordinary (or disproportionate) means of treatment. The core point of this distinction is that ordinary means of treatment typically are morally obligatory for patients: "A person has a moral obligation to use ordinary or proportionate means of preserving his or her life" (*ERD*, 56). In contrast, extraordinary means of treatment are morally optional for patients: "A person may forgo extraordinary or disproportionate means of preserving life" (*ERD*, 57).

A closer look at this distinction reveals important insights in Catholic conservative teaching that can be construed, surprisingly to many, as progressive. An obvious question deals with the meaning of ordinary and extraordinary. The core distinction revolves around the use of burden and benefit language. The first insight is that burden/benefit align very differently to categorize what is ordinary or extraordinary. Basically, for a treatment to be obligatory (ordinary) requires a much higher bar than for a treatment to be optional (extraordinary). This contrast revolves around the words *and/or* as reflecting a higher/lower bar. Here are the teachings of the *ERD*s. First, "[p]roportionate means are those that … offer a reasonable hope of benefit *and* do not entail an excessive burden …" (*ERD*, 56, emphasis added). Second, "[d]isproportionate means are those that … do not offer a reasonable hope of benefit *or* entail an excessive burden" (*ERD*, 57, emphasis added). In other words, for a treatment to be obligatory (ordinary), the interventions must offer a reasonable hope of benefit *AND* avoid an excessive burden. In contrast, for a treatment to be optional (extraordinary), the interventions either do not offer a reasonable hope of benefit *OR* entail an excessive burden. In sum, for an intervention to be obligatory a much higher moral justification is needed; for a treatment to be optional requires a much lower moral justification.

The second insight deals with the central role of the patient in ascertaining when a means of treatment is obligatory (ordinary) or optional (extraordinary). The *ERD*s emphasize that the moral distinction is a matter for the judgement of the patient. First, "[p]roportionate means are those that *in the judgment of the patient* offer a reasonable hope of benefit and do not entail an excessive burden …" (*ERD*, 56, emphasis added). Second, "[d]isproportionate means are those that *in the patient's judgment* do not offer a reasonable hope of benefit or entail an excessive burden …" (*ERD*, 57, emphasis added). This may be surprising to some who consider the determination of a treatment as optional or obligatory should be construed in a different manner (e.g., by the authority of Church teaching). It is unambiguous that the pivotal determination of a treatment to be morally optional or obligatory ultimately is the responsibility of the patient.

The third insight deals with another component that determines the distinction between optional (extraordinary) and obligatory (ordinary) treatment. In addition to the indispensable contribution of the prepositions *and/or*, and to the central role of the patient in discerning whether a treatment is morally optional or obligatory, there is a surprising but crucial role for the family and the community. The concept of burden is accompanied by the concept of expense, regarding both ordinary and extraordinary treatment. The famous burden/benefit analysis (that in medicine typically addresses the various aspects of treatment) is extended to include expense. In other words, when weighing the burden analysis, a legitimate expense component is involved, and the expense can be aligned with family or society. The *ERD*s are explicit about this inclusion of expense. First, "[p]roportionate means are those that in the judgment of the patient offer a reasonable hope of benefit and do not entail an excessive burden *or impose excessive expense on the family or the community*" (*ERD*, 56, emphasis added). Second, "[d]isproportionate means are those that in the patient's judgement do not offer a reasonable hope of benefit or entail an excessive burden, *or impose excessive expense on the family or the community*" (*ERD*, 57, emphasis added). In sum, when a patient seeks to determine whether a treatment may be morally optional or obligatory, the discernment about burden may be accompanied by a concern about expense, and the expense can pertain either to the family or to society at large. This is a breathtaking tenet of Catholic doctrine that guides all end-of-life care decisions about death and dying.

Having made these underlying clarifications, the practical issues related to death and dying in the Catholic tradition can be divided into two general groups: those dealing with embryos in maternal-fetal conflicts and those dealing with patients whose lives (young or old) are ending. Each group is discussed in turn.

12.5 Maternal-Fetal Conflicts

The discussion earlier to explain the Principle of Double Effect introduced a classic case of maternal-fetal conflict, that of a mother (with a pre-viable fetus) with cancer of the womb. There is considerable debate in secular bioethics about the human status of the embryo in its early stages of development, and Catholic bioethics continues to engage that discourse especially as science generates new insights into embryogenesis (Magill and Neaves 2009, pp. 23–32; Magill 2009, pp. 101–135). However, to grasp the intensity of the moral dilemma of maternal-fetal conflicts, we can assume that the human status of the fetus in question is not disputed. The four conditions of the Principle of Double Effect that are used to resolve this moral dilemma can be applied to similar cases of maternal-fetal conflict. A recent controversy presented a dilemma that confronts Catholic health care regularly, the situation where the placenta exacerbates a condition of pulmonary hypertension in the mother that causes a threat of her imminent death (Magill 2011, pp. 848–865).

In 2009, a patient in this circumstance at St. Joseph's Hospital and Medical Center in Phoenix, Arizona led to an evacuation of her womb with the accompanying death of her 11-week old fetus. The facility was Catholic and the 27-year old mother was a practicing Catholic with four children. After a surprise pregnancy, the mother's previous condition of pulmonary hypertension dramatically worsened—in the emergency room the diagnosis included severe pulmonary arterial hypertension, from which two other pathologies emerged, right-sided heart failure, and cardiogenic shock that can result in cardiac arrest. The mother's death was imminent and the patient could not be transferred to another hospital. After an Ethics Consultation

(Magill 2013b, pp. 761–774), the Ethics Committee of the Catholic facility supported a dilation and curettage (D&C) procedure that ended the pregnancy with the mother's consent. Almost a year later, the bishop of the Phoenix diocese deemed the procedure to be a direct abortion. Each side held its ground, leading to much subsequent disagreement. A brief account of the stance of each side can help to clarify how the traditional Principle of Double Effect can be applied.

The opposing arguments of each side called upon the *ERD*s for support. On the one hand, *ERD* 45 states: "Abortion (that is the directly intended termination of pregnancy before viability or the directly intended destruction of a viable fetus) is never permitted. Every procedure whose sole immediate effect is the termination of pregnancy before viability is an abortion." This teaching appears to have been the basis of the decision of the bishop to condemn the intervention (that led to the death of the fetus) as an intended abortion. On the other hand, *ERD* 47 states: "Operations, treatments, and medications that have as their direct purpose the cure of a proportionately serious pathological condition of a pregnant woman are permitted when they cannot be safely postponed until after the unborn child is viable, even if they will result in the death of the unborn child." This teaching appears to have been the basis of the decision of the Ethics Committee to justify the intervention that saved the mother's life while foreseeing but not intending the death of the fetus.

To understand how the case might be resolved morally, it is important to recognize the role of the placenta as a shared organ of the mother and the fetus during pregnancy, though after birth it is unnecessary and discarded. The imminent threat of the mother's death (which would have inevitably have led to the death of the pre-viable fetus) was caused by the placenta exacerbating the mother's condition of pulmonary hypertension. The offending organ so to speak was the placenta, not the fetus. Even if, as may have been the case, the fetus was dead in the uterus given the circumstances of the pregnancy, the placenta could continue to function and present a threat to the mother's life. To resolve the medical crisis, the placenta had to be removed. However, to remove the placenta required evacuating the uterus including the amniotic membranes that contained the fetus.

The traditional Principle of Double Effect can be used to justify the decision of the Ethics Committee. To appreciate how the principle can contribute to the case, a brief consideration of each of its conditions is necessary. It can be helpful to bear in mind how this Principle pertains to the case of the mother (with a pre-viable fetus) with cancer of the womb.

The first condition of the Principle is that the physical action itself should be either morally good or neutral. In the case of the cancerous uterus the hysterectomy is the act under consideration. In the hypertension case, the act under consideration is the removal of the placenta through evacuation of the womb. In each case, the offending organ (the cancerous womb or the placenta) must be distinguished from the fetus. Hence, the first condition of the Principle is met based on the legitimacy of removing the offending organ that threatens the mother's life. The second condition of the Principle is that the bad effect (the death of the fetus) does not cause the good effect. In the case of the cancerous uterus, removing the fetus via abortion would not resolve the underlying pathology of cancer. In the hypertension case, even if the fetus was dead in the womb, the placenta would continue as the offending organ. In each case, the bad effect (the death of the fetus) did not cause the good effect (saving the mother's life). The third condition of the Principle is that the good action is what is intended. In the case of the cancerous uterus, the removal of the womb as the offending organ is what is intended. In the hypertension case, the removal of the placenta as the offending organ is what is intended. In each case, the death of the pre-viable fetus is unavoidable, foreseen, but not intended. The fourth condition of the Principle is that there is proportion between the good action (saving the life of the mother) and the bad side-effect (the death of the fetus). In both the case of the cancerous uterus and the hypertension case, the interventions are necessary to save the one life that can be saved. In each case, to do nothing would lead to the death of both the mother and the fetus. The obligation to rescue the mother when otherwise both lives would be lost demonstrates the proportionality of the intervention.

In sum, the four conditions of the Principle of Double Effect pertain as much to the hypertension case as to the classical case of the cancerous pregnant uterus. Of course, the foreseen but unintended death of the fetus in each case is tragic but inevitable. These complex cases shed light on the contribution of the Principle of Double Effect in the Catholic tradition to resolve various forms of maternal/fetal conflict. The Principle also is used in the more frequent (but no less soul-searching) moral dilemmas that arise regarding patients approaching the end of life.

12.6 Patients Approaching the End of Life

When patients receive medical treatment as they approach the end of life, perhaps the most pivotal moral decision deals with what is referred to as futile treatment. The debate over futile treatment in the Catholic tradition raises profound concerns about the medicalization of death. Although Catholicism clearly repudiates the continuation of what can be reasonably construed to be futile treatment, care for patients must continue to their death—accepting the reality of futile treatment is very different from any construal of futile care. The discussion in secular bioethics understands medical futility as declining medically ineffective treatment that would not offer a patient any significant benefit or would be contrary to generally accepted health care standards (National Conference of Commissioners on Uniform State Laws 1994, Section 7f). In terms of the well-recognized principles of bioethics, futility helps to clarify these points. The principle of autonomy of the patient does not oblige physicians to provide all treatments. The principle of beneficence recognizes there are times when treatments provide no benefit to a patient. The principle of non-maleficence highlights that ineffective treatment can cause harm to a patient by increasing suffering or delaying comfort care. The principle of justice sheds light on the avoidable costs that can be accrued to society when ineffective or wasteful treatment is provided to a patient. These secular perspectives about futile treatment in health care are recognized by the Catholic tradition in its recognition (as mentioned above) that "the duty to preserve life is not absolute" (*ERD*, Section Five, Introduction), and that in "the use of life-sustaining technology … two extremes are avoided: … an insistence on useless or burdensome technology … and, … the withdrawal of technology with the intention of causing death" (*ERD*, Section Five, Introduction).

The connection between avoiding futile treatment and avoiding causing death leads to the next point, withdrawing life-sustaining interventions like medically assisted feeding (nutrition and hydration), such as with patients in a persistent vegetative state (PVS). The focus on PVS patients in the *ERD*s resulted from the famous case of Terri Schiavo who died after 15 years of medically assisted nutrition and hydration. In 1990, Terri collapsed with a heart attack at home, depriving her brain of oxygen and leading to a PVS condition. The debate over withdrawing her life-sustaining treatment revolved around a disagreement between her husband who was guardian and her parents. Her husband argued that Terri would not have wanted to live in the manner offered via PVS treatment and sought withdrawal of the artificial feeding, and her parents contested the PVS diagnosis, seeking to keep her alive. After many court interventions at different levels, the final court decision permitted withdrawal and Terri died after 2 weeks on March 31, 2005.

The high-profile nature of this unfortunate case led the Catholic Bishops to revise the earlier version of the ERDs to provide guidance to Catholic health care in the treatment of PVS patients. On the one hand, Catholicism acknowledges that "while medically assisted nutrition and hydration are not morally obligatory in certain cases, these forms of basic care should in principle be provided to all patients who need them, including patients diagnosed as being in a persistent vegetative state (PVS)" (*ERD*, Section Five, Introduction). On the other hand, Catholicism accepts that "medically assisted nutrition and hydration become *morally optional* when they cannot reasonably be expected to prolong life or when they would be excessively burdensome for the patient or would cause significant physical discomfort" (*ERD*, p. 58, emphasis added). These statements indicate the fine balancing act that the Catholic tradition seeks to establish, expecting in principle to provide such treatments, but recognizing in practice the situations when the treatment becomes too burdensome and hence constitutes an extraordinary means that is not obligatory (as explained earlier). In this analysis, it should be recalled that the decision about ordinary (obligatory) and extraordinary (optional) means belongs to the patient or the patient's legitimate surrogate.

When life-sustaining treatment is futile and legitimately withdrawn from a patient, the issues of palliative and hospice care arise. In previous generations, there was a distinction between these noble services as somewhat distinct, whereas they are now seen in an integrated manner. The integrative approach initiates palliative care when a life-limiting diagnosis occurs. With treatment and palliation working together, as treatment becomes increasingly less effective palliative measures increase in scope, eventually leading to hospice interventions to maximize end-of-life care.

The Catholic tradition is explicit about the legitimacy of palliative and hospice care, including scenarios in which medications may shorten life, though modern medicine make that less likely today than in previous times: "Medicines capable of alleviating or suppressing pain may be given to a dying person, even if this therapy may indirectly shorten the person's life so long as the intent is not to hasten death" (*ERD*, 61). The importance of intention here reiterates the stance mentioned above that we must avoid "the withdrawal of technology with the intention of causing death" (*ERD*, Part Five, Introduction). The core concern here is to be abundantly clear that in the Catholic tradition "suicide and euthanasia are never morally acceptable options" (*ERD*, Part Five, Introduction). By euthanasia is meant "an action or an omission which of itself or by intention causes death, in order that all suffering may be in this way eliminated" (*ERD*, Part Five, Introduction).

The concept of suicide includes the prohibition of assisted suicide. This stance is especially salient in the face of increasing legislation in various states to permit assisted suicide in the U.S. Recently, on October 5, 2016 Governor Jerry Brown of California signed into law the *California End of Life Option* (approved by the State Assembly and Senate in 2015). This occurred in the wake of the assisted suicide of 29-year old with brain cancer, Brittany Maynard from California, who moved to Oregon to access legalized assisted suicide in 2014. States in the U.S. currently with assisted-suicide legislation (though adopted in different ways) include California, Montana, Oregon, Vermont, and Washington, and there are many others with aid-in-dying legislation under consideration. The provision of and affordable access to palliative and hospice care programs is increasingly becoming an indispensable platform for Catholic health care to resist legislatures that seek to permit assisted suicide.

Unfortunately, when managing end-of-life pain, there are extreme situations where the suffering of a patient cannot be effectively managed by medications in palliative or hospice care. In such cases, there is a last resort intervention that is permissible in the Catholic tradition. This is referred to as terminal sedation in the sense that a patient is rendered unconscious through sedation knowing that the patient will die shortly thereafter. Catholic teaching articulates the issue in this manner. Again, the general context of this stance is "… the duty to preserve life is not absolute …" (*ERD*, Part Five, Introduction). The specific teaching is as follows: "Patients should be kept as free of pain as possible … Since a person has the right to prepare for his or her death while fully conscious, he or she should not be deprived of consciousness without a compelling reason" (ERD, 61). The clear implication here is that patients may be deprived of consciousness (such as in terminal sedation) if there is a compelling reason, that unconsciousness is the only way to effectively treat the patient who is suffering from otherwise unmanageable pain.

Once again, the long shadow of the Principle of Double Effect can be seen here. The point is to address the pain of the patient, not to intend the patient's death. The principle's four conditions are met. The first condition is that the act itself is either morally good or neutral—in this case, rendering a patient unconscious (the frequency of this occurrence in surgery indicates that it is inherently not a bad action). The second condition is that the bad effect (the eventual death of the patient) is not caused by the good act (for example, many patients in scenarios that are not at the end of their lives are rendered unconscious but recover from their underlying pathology). The third condition is that the good action must be what is intended—in this case, resolving the patient's suffering from otherwise unmanageable pain. The fourth condition is that there is proportion between the good action (being rendered unconscious to address pain) and the bad side-effect (inevitable death)—in this case the patient is at the end of life and death is inevitably foreseen but certainly not intended. This stance of the Catholic tradition on terminal sedation, though perhaps surprising to some, is a fine illustration of the flexibility and range of the Principle of Double Effect.

12.7 After Death Dilemmas

There are two fascinating situations that occur after death for which Catholicism appears to adopt a surprising stance: organ donation after cardiac death (DCD) and posthumous pregnancy. A brief word of each can shed light on the astuteness of the Catholic tradition.

Organ donation after cardiac death is different from donation after brain death has been diagnosed. There is general agreement that brain death indicates a reliable measure for the end of someone's life, though discussion continues on this, but DCD presents an interesting ethical conundrum. DCD typically occurs when a patient at the end of life opts to donate organs after death. These patients are usually on life support and give consent to the DCD procedure, as follows. They are taken to surgery where life-support measures are withdrawn. Many facilities adopt the so-called Pittsburgh Protocol that waits for 2 min after death is determined (using cardio-circulatory criteria). The rationale for this timeline is that after the 2 min period there is no evidence of a patient auto-resuscitating. Then a different surgical team enters to remove the organs. But here is the dilemma. Technically, after 2 min the patient could be artificially resuscitated (this should not occur of course because the patient's consent for the DCD process would forbid doing so). If resuscitated, there would likely be extensive brain damage, but the patient might technically be construed as being alive. Does this mean that organ removal (just 2 min after the heart stops beating) constitutes killing? Of course, the individual will certainly be dead after the removal of major organs! Or is DCD a morally acceptable practice, despite the above conundrum?

Currently, Catholic teaching and practice accepts DCD as morally permissible, despite the conundrum. Here is a summary of Catholic teaching that relates to this problem. First, organ donation in general is construed as a moral good: "Catholic health care institutions should encourage and provide the means whereby those who wish to do so may arrange for the donation of their organs and bodily tissue, for ethically legitimate purposes, so that they may be used for donation and research after death" (ERD, 63). Second, Catholic teaching recognizes that when the exact point of death occurs (in the sense of the soul leaving the body) is impossible to know. That is why Catholicism accepts the time of death based on reliable scientific expertise: "The determination of death should be made by the physician or competent medical authority in accordance with responsible and commonly accepted scientific criteria" (*ERD*, 62). Hence, Catholicism remains open to DCD procedures despite the above conundrum, provided cardio-circulatory criteria for death remain accepted by clinical experts. Third, given this stance, Catholic teaching insists on separating the personnel who determine death from those who retrieve the organs to avoid professional conflict, reflecting the current standard and practice in health care: "to prevent any conflict of interest, the physician who determines the death should not be a member of the transplant team" (*ERD*, 64).

The second conundrum deals with posthumous pregnancy. If a pregnant woman dies, is there any obligation to maintain the fetus alive in the dead mother's womb using medical technology? Two interesting cases occurred in recent years. In November 2013, Marlise Munoz was 14 weeks pregnant when she collapsed and was deprived of oxygen for 1 h. She was taken to a hospital in Fort Worth, Texas, and placed on life support, even though there was a determination of brain death. The hospital insisted it was obliged by a Texas statute that forbade withdrawing life-sustaining measures from a pregnant woman who had died—the hospital's legal team sought to avoid a charge of criminal homicide related to causing the death of a fetus. In January 2014, the husband sued to release the body for burial. The judge ordered the removal of life-support based on the rationale that Texas law did not apply to the case because the patient was legally dead (being brain dead). The hospital complied and the fetus died. A different outcome occurred in December 2013 with Robyn Benson who became unconscious from a fatal brain hemorrhage, and was declared dead in a general hospital in British Columbia. Her husband Dylan asked that she remain on life-support to let the fetus develop sufficiently for delivery. The hospital agreed, baby Ivon Cohen was delivered at 28 weeks, and the dead mother was removed from the technology in February 2014.

This conundrum of a posthumous pregnancy (being brain dead and pregnant) occurs with sufficient frequency as to have led to a variety of legal statutes across the U.S. with varying requirements in these situations. The conundrum for the Catholic tradition is whether its pro-life stance requires Catholic facilities to use of life-sustaining measures on a brain-dead mother until the fetus is sufficiently developed to be delivered. To resolve this conundrum requires returning to the core distinction between ordinary (obligatory) and extraordinary (optional) means. Despite the pro-life stance of Catholicism to protect the unborn, we should be mindful of the core teaching of Catholicism that "the duty to preserve life is not absolute" (*ERD*, Part Five, Introduction). Hence, in a dilemma like posthumous pregnancy the moral decision must be based on the balance between benefit and burden (including expense as highlighted earlier) implied by the distinction between ordinary and extraordinary means. As a result, it can be reasonably inferred that given the burden/expense incurred by the hospital and society (there is no burden for the dead mother) of using life-sustaining measures for a long period of time, the intervention should be construed as extraordinary, and hence not obligatory. In other words, if a family and health care provider and insurer can accept the burden/expense of life-sustaining measures, morally they may be adopted (as extraordinary means), but they are not morally required to be adopted (which could only pertain if they were ordinary means).

12.8 Conclusion

The previous sections have moved from the theoretical realm to consider several practical dilemmas that ethically engage issues surrounding the medicalization of death. The focus on the Catholic tradition has shown how a religious denomination relies strenuously upon reasoning and science to guide its official teaching in general and to make moral decisions in practice. The flexibility and range of Catholic morality can offer astute guidance not only to those who follow this tradition but also to those with different faith perspectives (or none) when they encounter the heart-wrenching dilemmas that medical technology presents around death and dying.

References

Magill, Gerard. 2009. Using excess IVF blastocysts for embryonic stem cell research: Developing ethical doctrine, secular and religious. *Hofstra Law Journal* 37: 101–135.

———. 2011. Threat of imminent death in pregnancy: A role for double-effect reasoning. *Theological Studies* 72 (2011): 848–878.

———. 2013a. Organizational ethics. In *Contemporary Catholic health care ethics*, ed. David F. Kelly, Gerard Magill, and Henk ten Have, 2nd ed., 271–290. Washington, DC: Georgetown University Press.

———. 2013b. Quality in ethics consultations. *Medicine, Health Care and Philosophy* 16 (4): 761–774.

———. 2014. Bioethics in Catholicism. In *Handbook of global bioethics*, ed. Henk A.M.J. ten Have and Bert Gordijn, 356–373. New York: Springer.

———. 2015. *Religious morality of John Henry Newman: Hermeneutics of the imagination*. New York: Springer.

———. 2017. Complicity of Catholic healthcare institutions with immoral laws. In *Contemporary controversies in Catholic bioethics*, ed. Jason Eberle, 523–536. New York: Springer.

Magill, Gerard, and William B. Neaves. 2009. Ontological and ethical implications of direct nuclear reprogramming. *Kennedy Institute of Ethics Journal* 19 (1): 23–32.

Magill, Gerard, and Lawrence Prybil. 2011. Board oversight of community benefit: An ethical imperative. *Kennedy Institute of Ethics Journal* 21 (1): 25–50.

National Conference of Commissioners on Uniform State Laws. 1994. *Uniform health-care decisions act*.

United States Conference of Catholic Bishops (USCCB). 2009. *Ethical & religious directives for Catholic health care services*. 5th ed. Washington, DC: USCCB. The references remain the same in the 6th ed., published in 2018.

Chapter 13
Islamic Perspectives on Clinical Intervention Near the End of Life: We Can but Must We?

Aasim I. Padela and Omar Qureshi

Abstract The ever-increasing technological advances of modern medicine have increased physicians' capacity to carry out a wide array of clinical interventions near the end of life. These new procedures have resulted in new "types" of living where a patient's cognitive functions are severely diminished although many physiological functions remain active. In this biomedical context, patients, surrogate decision-makers, and clinicians all struggle with decisions about what clinical interventions to pursue and when therapeutic intent should be replaced with palliative goals of care. For some patients and clinicians, religious teachings about the duty to seek medical care and the care of the dying offer ethical guidance when faced with such choices. Accordingly, this paper argues that traditional Sunni Islamic ethico-legal views on the obligation to seek medical care and Islamic theological concepts of human dignity (*karāmah*) and inviolability (*ḥurmah*) provide the ethical grounds for non-intervention at the end of life and can help calibrate goals of care discussions for Muslim patients. In closing, the paper highlights the pressing need to develop a holistic ethics of healthcare of the dying from an Islamic perspective that brings together multiple genres of the Islamic intellectual tradition so that it can meet the needs of the patients, clinicians, and Muslim religious leaders interacting with the healthcare system.

First published through Springer Nature in 2016. In *Medicine, Health Care, and Philosophy*. doi:10.1007/s11019-016-9729-y. Thanks to Springer for the right to reuse.

A. I. Padela
The University of Chicago, Chicago, IL, USA
e-mail: apadela@uchicago.edu

O. Qureshi (✉)
Zaytuna College, Berkeley, CA, USA
e-mail: oqureshi@zaytuna.edu

T. D. Knepper et al. (eds.), *Death and Dying*, Comparative Philosophy of Religion 2, https://doi.org/10.1007/978-3-030-19300-3_13

13.1 Introduction

Novel technological advances in science and medicine provide physicians with a greater number of tools and an increased capacity to restore and supplant the functions of human organs. In light of these newfound capabilities, much ink is being spent discussing the ethics of end-of-life healthcare. On one hand, the technical powers of biomedicine are clearly life-saving. Motor-vehicle accident victims who present with intracranial hemorrhage, for example, now have a chance at survival and at returning to normal functioning due to surgical and technological innovations. Yet, at the same time, increased scientific knowledge and technical prowess can result in outcomes that are ambiguous where patients linger in minimally conscious states within nursing facilities. Dramatic technical advances have also enabled biomedicine to socio-culturally construct (or label new) "states" of life and "types" of death, including persistent vegetative state, brain death, and the donor after cardiac death. These new constructions can and usually do obscure previously clear distinctions between the living and the dead. These physiological states and types of life and death, whether resulting from prior clinical interventions or servicing future ones, are sites of moral entanglement and ethical dilemmas.

In this clinical context, religious traditions may help clinicians, patients, and other healthcare stakeholders to perform the moral calculus related to clinical intervention near the end-of-life by defining human life and describing a life worth medically maintaining. Religious ethical frameworks also assist in healthcare decision-making by demonstrating how scriptural source-texts and religious teachings can be used to evaluate the merits of medical intervention in a rapidly changing biomedical landscape. Accordingly, this paper addresses the moral dimensions of medical practice near the end of life from an Islamic perspective.

Like other communities, Muslim patients, clinicians, and Islamic scholars grapple with questions about the ethical obligations of providers and families during end-of-life healthcare, and they struggle with the clinical uncertainties related to the shifting borders between life and death. For example, a qualitative study of immigrant Muslim physicians in the United States found there to be tensions between their faith and end-of-life care as participants felt "(withdrawal) would be against the religion" (Padela et al. 2008, p. 367). Confirming that American Muslim physicians are challenged by of end-of-life clinical decision-making, a recent national survey reported that 70% of respondents felt withdrawing life-sustaining treatment caused greater psychological distress than withholding it. The survey also identified additional sources of tension between Islamic values and contemporary end-of-life care, as nearly 50% of respondents were unsure whether brain death signifies true death according to Islam or whether Islam permits feeding tubes to be withdrawn (Padela unpublished data). Even in Saudi Arabia where brain death is considered death by law, a longitudinal study of brain-dead patients found that terminal extubation took place only among a small minority, and that Muslim families were conflicted about limiting clinical interventions for brain-dead family members (Khalid et al. 2013).

Brain death is a source of much controversy within the Islamic juridical community and debates over brain death illustrate the struggle religious leaders face in providing ethical guidance for end-of-life healthcare. Over the past several decades, prominent Islamic juridical councils across the globe have taken up the question of whether brain death can be considered ontological death and whether the clinical criteria suffice for legal death within Islam. Some accept brain death as legal death in Islam, others consider it to be a dying state, whereas a minority reject brain death as death (Farah and Al-Kurdi 2006; Padela et al. 2011, 2013; Padela and Basser 2012; Sachedina 2017; Tavakkoli 2008; Moosa 1999; Ebrahim 1998). Consequently, while the former two camps find it ethico-legally permissible to withdraw life support from brain dead patients, the latter does not (Miller 2015; Padela et al. 2013). As the science around brain death is further clarified and clinical practice guidelines for assessing brain death continue to evolve, both Muslim clinicians and ethicists have called for revisiting Islamic perspectives on brain death (Qazi et al. 2013; Rady and Verheijde 2013; Padela et al. 2011; Miller et al. 2014; Hamdy 2013; Moosa 1999; Bedir and Aksoy 2011).

These studies illustrate that healthcare at the end of life is fraught with multiple ethical challenges for Muslim patients, providers, and religious leaders. This paper aims to provide some guidance and advances to the bioethical discourse regarding healthcare for the imminently dying by retrieving classical Sunni juridical perspectives on whether Muslims are obligated to seek medical care and by discussing Islamic theological correlates for human dignity and inviolability. Our review of Islamic legal opinions will demonstrate how clinical intervention is, in general, not morally obligated upon Muslims, and our discussion of human dignity and inviolability will suggest that these constructs provide a counter-weight to continued clinical interventions near the end of life. Finally, the paper will close by commenting on the urgent need for a theologically rooted, holistic bioethics of caring for the dying from an Islamic perspective in order to better serve Muslim physicians and patients.

13.2 The Moral Status of Seeking Medical Care in Sunni Islam

13.2.1 Research Methods and Terms of Engagement with Islamic Law

Before proceeding to describe Islamic ethico-legal perspectives on seeking medical care, a few limitations must be presented, and a few critical terms need to be defined. Islam is divided into two major theological sects: Sunni and Shia, with approximately 85% of Muslims considering themselves to be Sunni (Pew Research Center 2009). While Sunni and Shia theology share much in common, they differ on who they consider as authorities for scriptural transmission and interpretation as well as on the role of reason in determining moral obligations. Accordingly, each sect has its own distinctive moral theology (*uṣūl al-fiqh*).[1] A *madhhab*, or a school of law, in the Islamic legal tradition consists of a body of legal opinions and hermeneutics developed by the eponymous founder of the school. The term applies to the founder's legal opinions as well as the opinion of jurists who subscribed to the hermeneutic of the school. This paper presents arguments from the Sunni schools of law because they derive their ethico-legal positions using, more or less, the same scriptural sources and tools, and they mutually recognize each other's truth claims (Kamali 2003). The four extant schools of law within Sunni Islam are Ḥanafī, Mālikī, Shafiʿī, and Ḥanbalī.

Islamic moral theology (*uṣūl al-fiqh*) stems primarily from two scriptural sources: the Qur'ān and the normative practice of the Prophet Muḥammad (*sunna*). Both of these sources are a part of the same revelatory transmission and are thus classified as divine communication (*waḥy*) (Doi 1984). Using these two sources as the fountainheads for moral duties, Islamic scholars have elaborated a science, an Islamic moral theology—*uṣūl al-fiqh*—by which to assess actions along a moral gradient from obligatory to forbidden (Fadel 2008).[2] An assessment of this type is termed *ḥukm taklīfī*, and it links human action to expected afterlife ramifications—God's reward, punishment, or indifference (Kamali 2003). Importantly, the moral status of actions is determined by examining the posited afterlife ramifications of an act through study of scriptural source-texts.

In this paper we present the *aḥkām*, ethico-legal rulings, about the moral status of seeking medical treatment from the four schools of Sunni law. In presenting these stances we draw on positions of authoritative jurists as recorded in standard legal manuals and instructional *fatāwā* compendia used within seminaries for teaching legal theory and sources that are representative of the major verdicts of a particular school. This selection is necessary because within any *madhhab,* one may encounter multiple viable legal positions on any given issue, all of these positions having been derived by using the particular ethico-legal theory and constructs of that school. Given this diversity, each school has developed a framework that provides a hierarchy of authorities and a categorization schema that allows for navigating the multiple opinions. For example, the term *al-aẓhar* refers to the strongest position among the various legal positions held by al-Shafiʿī on a particular issue, whereas the term *al-aṣaḥ* refers to the "most correct" position according to the jurists associated with the Shafiʿī but it is not the position of al-Shafiʿī himself. In this study, we examine

[1] We adopt Prof. Mohamed Fadel's usage of the English term moral theology to refer to the Islamic science of *uṣūl al-fiqh* (Fadel 2008). As Prof. Fadel notes, insofar as *uṣūl al-fiqh* is concerned with the scriptural sources of moral obligation, the processes of moral assessment, and moral epistemology, it is a moral science. And since *uṣūl al-fiqh* is primarily concerned with how God judges human acts and strives to reach the truth regarding moral propositions, it is a theological discipline. Consequently, the mapping of terms is apropos even if not precise. I use the terms "Islamic ethico-legal tradition" and "Islamic law" to refer to the notions of *fiqh* and *aḥkām taklīfiyya* interchangeably.

[2] This gradient ranges from obligatory (*farḍ* or *wājib)* to recommended (*mandūb* or *mustaḥab)* to permitted (*mubāḥ*) to discouraged (*makrūḥ*) and, finally, to prohibited (*ḥarām*).

the authoritative works of each school and work with the positions that jurists of the school have identified as the strongest.

While our purposive sampling provides insight into a normative Islamic ethico-legal perspective and represents the prevailing position within a particular legal school, we acknowledge that secondary and non-dominant positions within these schools merit detailed study. Indeed, the diversity of positions within a school allows for jurists to use their discretion for penning *fatāwā* that remain within the bounds of a school's scholarly lineage yet at the same time attend to the contextual factors that weigh upon the one seeking the *fatwa*. Studying such *fatāwā* can yield insights into the creativity employed by jurists when confronted by ethical challenges accompanying scientific advancements. Yet, because *fatāwā* may prioritize contingencies and adopt minor positions from within (or even from outside of) the legal schools, *fatāwā* can represent exceptions to the rule. Because we desired to focus on the "rule" and not exceptions or ethico-legal innovations, we restrict our discussion to *fatāwā* compendia used for ethico-legal instruction and do not venture into modern *fatāwā* collections of jurists operating outside of the school structure.[3]

Finally, it bears mention that modernity has challenged notions of Islamic normativity and that the seminary, as well as *madhhab*, authority is hotly contested in the contemporary period. The *uṣulī* approach to ethico-legal judgment utilized by many seminarians is deemed as outmoded by scholars and new approaches such as *fiqh al-aqalliyāt* (jurisprudence for Muslim minorities) or embellishments of different genres of Islamic law such as the *maqaṣid al-shar'iah*, are offered as alternative frameworks for deriving ethico-legal injunctions (Auda 2008; Attia 2007; 'Alwānī and Shamis 2010). While we acknowledge these "newer" approaches to deriving Islamic ethico-legal stances and the reasoned critique of the legal school paradigm, the schools of law remain an authoritative source that informs the new approaches themselves as well as for present-day Muslims (both jurists and laity) around the world as they furnish the inherited canon. As such we hold that they are a key source to study when constructing an Islamic bioethics.

It also bears mention that many state and civic organizations as well as transnational bodies routinely convene juridical councils to offer Islamic legal guidance. These modern councils comprise jurists of varied backgrounds and legal persuasions, and operate on the basis of collective *ijtihād (jama'ī)*, which allows for drawing eclectically upon the conventions and constructs of the schools of law (Karman 2011). The resolutions of these councils provide critical insight into how jurists balance classical positions with modern contexts. Since the ethico-legal methodology operating in the deliberation chamber is ambiguous, and since the linkage between the resolutions and the schools of law is tenuous, however, the judgments of such councils do not suffice as primary sources for the present study. Rather, pertinent resolutions are mentioned in a complementary fashion below.

[3] For a classification schema of the different types of *fatāwā*, see Skovgaard-Petersen (2015), and for insights into legal manuals used for seminary instruction in Islamic law, see Fadel (1996).

13.2.2 The Scriptural Source-Texts

The reader will benefit from a description of the key scriptural source-texts at the heart of the debate on whether seeking medical care is obligated in Islam. The Qur'an unequivocally ascribes healing as an act of God as it quotes the Prophet Abraham saying, "And when I am ill, it is He [God] who cures me" (Ali 1999; 26:80). Statements from the Prophet Muhammad also relate that illness and cure are from God (see below). At the same time the Prophet described a group of people who will enter paradise without reckoning, saying that "they have never allowed themselves to be treated by cauterization…rather, they have put their reliance in God alone" (al-Bukhari 2002, 1610). Importantly, the Prophet described the reward for avoiding cauterization, a common medical treatment at the time, as entry into paradise. This statement creates tension with other statements of Prophet which instruct believers to pursue medical care. For example one narration states, "seek medical treatment, for except for senility, God has not created an illness except that He also created its cure" (al-Sijistānī 2009, 6:5). These statements suggest that Muslims are encouraged to seek medical treatment, but they also have a reward for abstaining. Against the backdrop of these scriptural sources, the prevailing opinion within each of the Sunni schools of law is described below.

13.2.3 The Ḥanafī School of Law

According to the Ḥanafī school, seeking medical treatment is not obligatory even if one dies because of this non-action. This position is expressed in several legal manuals (*mutūn*) such as *al-Mukhtār* of al-Mawṣilī (d. 683/1284) and *Multaqā al-Abḥur* of Ibrāhīm al-Ḥalabī (d. 956/1550). For example, al-Mawṣilī's *al-Ikhtiyār*, his commentary on *al-Mukhtār*, and Dāmād Afandī's (d. 1078/1667) *Majmaʿ al-Anhur*, a commentary on *Multaqā al-abḥur*, recall the ruling that that there is no sin upon the one who does not seek medical treatment, and then clarify that this determination has been made "because there is no certainty that this treatment will cure him and it is possible that he will become well without treatment" (al-Mawṣilī, 2:409–410; Dāmād Afandī 1910, 2:525). Prominent *fatāwā* collections within the school also corroborate this position. Illustratively, *al-Fatāwā al-Hindiyyah* cites *Fatāwā Qādīkhān* and states that if a doctor tells the patient that he needs a certain treatment, and the patient refuses the treatment and subsequently dies, the patient has not sinned (Niẓām 2009). The rationale underlying this position is stated in several *taʿlīl* works (monographs that detail the evidence and rationale behind ethico-legal rulings). Badr al-Dīn al-Simāwī's (d. 823/1420) *Jāmiʿ al-fuṣūlayn* expands on this point. The text mentions that the removal of harm by action can be either certain (*maqtūʿun bihi*) or probable (*maẓnūn*) or doubtful (*mawhūm*). Eating and drinking to relieve hunger and thirst (both representing harms) are actions that lead to the certain removal of hunger and thirst; however, medical treatment is from the second (probable) category, and therefore refusal is not sinful (al-Simāwī 1882). Importantly al-Simāwī mentions a possible exception to this rule, stating that if an individual knows by personal experience that a specific treatment will *certainly* remove the harm caused by disease, then for this person *that particular treatment* may be obligatory to use. Given that 100% clinical efficacy is rare, the potential zone of moral obligation to seek medical treatment appears to be small according to the Ḥanafī construct.

13.2.4 The Shafiʿī School of Law

The foremost authorities for legal opinions in the Shafiʿī school of law are Ibn Ḥajar al-Haytamī (d. 974/1566–1567) and MuḤammad ibn AḤmad al-Ramlī (d. 1004/1596), whose legal opinions are found in their commentaries on the school's central legal text, *Minhāj al-Ṭālibīn* by al-Nawawī (d. 676/1277). These authorities state the default ruling in the Shafiʿī school on seeking medical treatment is that it is a recommended but non-obligatory act. For example, al-Haytamī comments,

> Seeking medical treatment is recommended based on the rigorously authenticated report (of the Prophet Muhammad), "Seek medical treatment. For, except for senility, God has not created an illness except that He also created its cure." And in another rigorously authenticated transmission it states, "God has not sent an illness except that He also sent its cure." If one avoided medical treatment trusting [in God], then it is a virtuous act (*fa huwa faḍīlah*). The author (al-Nawawī) stated this. al-Adhraʿī considered [a person who does not seek treatment] to be superior explaining that if a person's trust is strong then it is better for him to not [seek medical treatment] but if [a person's trust] is not [strong], then [seeking treatment] is better …. Qāḍī ʿĪyāḍ (another authority in the Shafi school) has transmitted that there is consensus (*ijmāʿ*) that seeking medical treatment is not obligatory. This [claim] is opposed by some scholars of our school holding that it is obligatory [to seek medical treatment] in the case of a person who had a wound which they feared would lead to death (*yukhāfu minhu al-talaf*). [The case of medical treatment being recommended] differs from it being obligatory such as in the case of swallowing wine when choking or to apply a dressing to the phlebotomy site because of the certainty of its benefit (*li tayaqqun nafʿihi*). (al-Shirwānī 1972, vol. 3, p. 182–183)

In this passage, al-Haytamī presents the default ruling in the Shafiʿī school by interpreting Prophetic directives to seek medicine as evidencing a recommendation to seek treatment. He admits that there are some scholars who consider seeking medical treatment to be an ethically obligated act, i.e. sinful if not performed, but notes that they elevate the moral status from recommendation to obligation only when leaving medical treatment would lead to death and there is certainty in the clinical efficacy of a particular medical treatment preventing death. According to al-Haythamī although seeking medical treatment is generally recommended, if the certainty of its benefit exists then recourse to medical treatment becomes obligatory. Further explaining why Shafiʿī jurists held seeking medical treatment to be recommended but not obligated, al-Ramlī explains "seeking treatment is not obligatory, contrary to [the case of] one compelled to eat from a corpse and [the case of one] washing down a morsel of food with wine, due to the lack of certainty (*al-qaṭʿ*) in it being effective, which is contrary to these two cases" (al-Ramlī 1967, vol. 3, p. 19).

Somewhat muddying the waters, however, is the fact that some Shafiʿī authorities deem the high probability (*al-ẓann al-ghālib*) of an illness occurring as sufficient to make the act of *not* utilizing clinical treatment sinful. Consequently, avoiding sickness becomes obligatory. For example, in the case of dry ablution (*tayammum*), Shafiʿī jurists state that if a physician informs a patient there is a high probability that using water will result in a person getting ill (*al-ghālib ḥusūl al-maraḍ*), then it is forbidden for one to use water and he must perform dry ablution instead (al-Shirwānī 1972). Importantly this example does not indicate a moral obligation to seek medical treatment; rather the obligation is to avoid causing further harm.

In summary, the dominant position of the Shafiʿī school is that seeking medical treatment to be a recommended act that becomes obligatory when clinical efficacy is certain or highly probable (*ghalabat al-ẓann*) or if leaving off treatment results in certain death. Indeed, the Shāfīʿ jurist-theologian Imam al-Ghazālī (d. 505/1111) held that seeking medical treatment is obligatory only when the cure is certain and the proposed treatment is life-saving (Ghaly 2010; Albar 2007).

13.2.5 The Mālikī School of Law

Although jurists of the Mālikī School have discussed the permissibility of medical treatments, they have, in general, not detailed whether such treatments can be deemed mandatory. The prominent Mālikī jurist al-Dardīr (d. 1201/1786) in *al-Sharḥ al-Ṣaghīr ʾilā Aqrab al-Masālik* confirms the ruling of permissibility and only mentions obligation saying "seeking medical treatment is permissible. It may be obligatory... [if] the treatment's benefit should be known through the science of medicine" (al-Dardīr and Ṣāwī 1972, vol. 4, p.770). Unfortunately, al-Dardīr does not elaborate on when the general permissibility of seeking treatment becomes obligatory, and Aḥmad al-Ṣāwī in his commentary on the text does not offer any further comment. However, other Mālikī jurists provide some insight in the assessment of obligation. They classify medical treatments into those whose efficacy medical experts are certain of, those with probable (*madhnūn*) clinical efficacy, and those whose efficacy has not been established (*mawhūm*). The ethico-legal rulings pertaining to seeking medical treatment are made on the basis of the posited clinical efficacy of treatment, or on the certainty of harm without treatment. For example, Shaykh MuḤammad al-Khadīm offers that clinical treatment becomes obligatory when not doing so will have a (certain) fatal outcome (al-Khadīm 2011).

13.2.6 The Ḥanbalī School of Law

Jurists of the Ḥanbalī school such as Ibn Mufliḥ (d. 763/1362) report the dominant position of the school being that seeking medical treatment is permissible but abstaining is superior. He states "seeking medical treatment is permissible, however not utilizing it is more meritorious. [Imam Aḥmad] unequivocally stated this. In al-Marwūdhī's transmission, [Imam Aḥmad] said, 'Treatment is a dispensation. Not seeking out treatment is a degree higher than it'" (Ibn MufliḤ al-Maqdisī 1996, vol. 2, p. 333). Ḥanbalī jurists give preference to the reward for a person to patiently bear the harm caused by the illness by interpreting the Prophetic statement to seek out medical treatment as general advice (*irshad*) and not a moral directive. An earlier jurist, Ibn Qudāma al-Maqdīsī (d. 620/1223), also advances this position, noting that where clinical efficacy is high and there are likely to be no detrimental side effects, "[I hold that] seeking such treatment would be permissible (only)" (Ibn Qudāmah 2007, vol. 2, p. 52). Although, it is reported that the Ḥanbali jurist Ibn Taymiyya (d. 728/1328) did hold that medical treatment is obligated when cure is certain and the proposed treatment is life-saving (Albar 2007; Ghaly 2010), the prevailing position of the Ḥanbalī school remains that seeking medical treatment is not an obligation.

In summary, the dominant position in the Ḥanafī, Malikī and Ḥanbalī schools is that seeking medical treatment is permissible but not obligatory, while Shafiʿī jurists hold seeking medical treatment to be a recommended act. All of the four schools of Sunni law regard that leaving medical treatment becomes sinful under exceptional circumstances and in the minority of cases. Ḥanafī jurists consider forgoing medical treatment, even if this non-action results in death, not to carry the weight of sin, while Shafiʿī and Malikī authorities suggest that Muslims would be considered to be sinning should they not seek medical treatment when the malady will cause death. The elevation of moral status from a permitted or recommended act to an obligatory act appears to hinge on the certainty regarding a fatal outcome without treatment and/or certainty (or in the case of some Shafiʿī juridical opinions, dominant probability) about the clinical efficacy of treatment in removing harm associated with the illness.

Before moving to discussing how these rulings can inform decision-making, we would like to address several criticisms of the inherited Islamic ethico-legal canon that undergird the above judgments. An argument could be advanced that these classical positions on seeking medical treatment are outdated and no longer applicable. The nature of healthcare, as well as the types of and capacities of clinical intervention are vastly different today than they were at the time these dominant positions were crystallized. Consequently, the present state of technology and medicine is one that could not have been imagined or foreseen by classical jurists, and a revision of their views is needed. Certainly, such an argument has a measure of truth in it because every bioethical framework requires updating as human knowledge increases, and religious hermeneutics are often reimagined in the light of newer scientific understandings of the world. At the same time the juridical positions outlined above appear to account for the deliverables of medical science and thereby fortify themselves, at least partially, against this critique. The conditions set forth by schools that make clinical treatment mandatory require scientific evidence. For example, the Ḥanafī school requires certainty that the intervention will remove illness-related harms, and although such certainty can reside in the patient more often than not, empirical claims are the basis for such certainty. The Shafiʿī view follows the same pattern grounding a moral obligation to seek treatment in the assessment of clinical efficacy or knowledge that leaving off treatment would result in death. Here too, clinical epidemiology and biostatics can deliver the answers. The Malikī position follows the same pattern as the Shafiʿī view. Only the Ḥanbalī view that medical treatment is never morally obligated distances itself from any consideration of the deliverables of medical science.

Modern *fiqh* academies seem to suffer from poorly accounting for modern scientific tools as well. For example, in 1992 the Organization of Islamic Cooperation sponsored Islamic Fiqh Academy, which includes Sunni jurists from all four schools of Islamic law as well as Shia jurists, met to revisit classical positions on seeking healthcare. After deliberating over the state of biomedicine today and reviewing classical stances, the council issued a resolution that largely coheres with the more classical positions of the Shafiʿī and Malikī views. They state that seeking modern medicine is obligatory when neglecting treatment may result in (i) the person's death, (ii) loss of an organ or disability, or (iii) if the illness is contagious and a harm to others (Islamic Fiqh Academy 2000). This more modern view added categories (ii) and (iii) to the classical stances. Notably however the resolution does not mention how clinical efficacy informs moral obligation, and it is unclear whether the juridical council delved into discussions about the epistemology of medical science or how biostatistics is used to assess clinical efficacy (Ghaly 2010). Consequently this "updated" ruling may also insufficiently account for modern biomedicine. Nonetheless, we believe revising the classical Islamic ethico-legal perspectives on the moral status of seeking medical treatment in light of the structure, epistemology, and tools of contemporary biomedicine is needed for constructing a holistic Islamic bioethics.

Similarly, one might argue that juridical perspectives from the past should not be treated as precedent because jurists of the past defined and deployed ethico-legal constructs based on the knowledge of their time. In other words as natural and social scientific knowledge has advanced, so too should the ways in which we understand the classical devices of Islamic law. This critique suggests that each epoch requires developing ethical constructs and frameworks that draw upon contemporary understandings of the world. Indeed as we come to understand the world around us more deeply, Islamic constructs that rely upon social assessments, e.g. *maṣlaḥa* (public and private benefit/interest), or natural world processes, e.g. *istiḥālah* (transformation), may need to be revised so that they can be appropriately used in Islamic ethico-legal deliberation (Padela et al. 2014). While we are sympathetic to this argument, a comprehensive methodology for updating Islamic ethico-legal theories and constructs has yet to be developed, and such an effort would require the massive mobilization of scholarly consensus in order to delineate which aspects of tradition are amenable to reimaging and which are not. Even if such a project were undertaken, the classical schools of law still offer, at a minimum, a credible starting-point for attending to the ethical challenges of modernity, and these opinions derived using traditional constructs and modes of reasoning maintain authoritative status within many institutions of religious learning today.

13.3 Relating the Moral Status of Seeking Medical Care to the Ethics of Clinical Intervention near the End-of-Life

To illustrate how these rulings might inform deliberations about clinical interventions let us consider an intervention aimed at relieving pain. Islamic jurists would first address the question of whether a Muslim is obligated to seek medical care for pain, or conversely whether the action of not seeking pain treatment carries sin. Scriptural evidences and legal precedents would be sought to support arguments for or against the obligation. Since the question of whether a Muslim is obligated to seek medical treatment is closely related to the question of whether a Muslim is obligated to seek pain relief (and when pain relief requires clinical intervention, it subsumes the question of seeking treatment), Islamic scriptures both suggest merit for an individual bearing pain, but they also command the removal of pain (when seen as a harm). There are multiple traditions from the Prophet Muhammad, for example, that attach reward to enduring pain and relate his enduring of severe pain at the end of his life (Mattson 2002; al-Bukhari 2002 Hadith 4428, 5640–5642); further, the Qur'an recounts the story of the Prophet Ayub (Job) to note that his forbearance with harms from illness and his reliance upon God in the face of these was rewarded (21:83–84). While such evidence funds the notion that forbearance with pain is rewarded and that forgoing therapy might be morally licit, the prophetic statement in Islam that there should be no harming or reciprocating of harm (*la ḍarar wa la ḍirar*) informs a cardinal maxim of Islamic law that harm must be removed (Ibn Mājah Hadith 2341).

In light of the question of the moral status of seeking clinical intervention in the case of pain (relief), a moral obligation would ensue according to some schools when the two aforementioned conditions for obligation are met: (i) whether the patient might be expected to die from the pain; and, (ii) whether there is sufficient research evidence that the proposed therapy will certainly or most probably remove the pain. Now since pain is not a physiological cause of death, the first condition is not met. However, since the strength of evidence for the efficacy of different modalities of pain relief is variable, the general ruling remains an open question. In moving from the general ruling (*ḥukm*) to a specific verdict (*fatwa*) regarding a specific intervention, jurists would need to examine whether the proposed intervention carries any harms during the course of removing the harm of pain since both clinical treatment and Islamic law at their cores aim at removing harm. Such an assessment could be made in light of the higher objectives of Islamic law since actions that undermine the higher objectives of Islamic law are considered harmful. For example, clinical therapies that contain substances deemed ritually impure or prohibited for consumption, e.g., porcine-derived hormones or other medicines, might be deemed as threats to the objective of preserving religion, while treatments that render a patient unconscious might threaten both the preservation of intellect and the preservation of religion because the unconscious individual loses decisional capacity that is required for discharging religious obligations. Finally, any other contextual factors relating to the patient and the proposed therapy would be considered prior to rendering a specific ethico-legal ruling. The results of this complex reasoning process might determine that an individual has a moral obligation to obtain the treatment, a moral obligation to refrain from treatment, or permissibility either way.

The critical point to note is that forgoing pain-control interventions remains an ethically justifiable option in light of the general rulings about the moral status of seeking medical care. From the Ḥanbalī perspective, abstaining from intervention would be preferred, while the Ḥanafī view would be that certainty about pain relief is required for the moral obligation and that forgoing treatment when the efficacy of the treatment is uncertain is permitted. Since the Shafiʿī and Malikī stances base moral obligation on whether death would ensue without treatment, pain-relief procedures from these two perspectives would appear to remain optional but not morally required since pain is not a proximate cause of death.

With respect to the ethico-legal argument for the non-obligated nature of clinical intervention near the end of life, a further comment regarding the conditions set forth by the jurists is necessary. Since some schools hold clinical treatment to be obligatory when non-treatment leads to death, and end-of-life clinical care necessarily deals with the impending death of a patient, one could argue that treatment at the end of life is obligatory. It is obligatory because impending death might be forestalled or the imminently dying state of the individual would be reversed. In the context of end-of-life care, however, it is important to recognize that many clinical interventions do not aim to avert death. Rather, interventions may be utilized for varied purposes including palliation or to maintain the viability of organs for possible donation. For example, an individual with oral cancer might seek the placement of a gastric feeding tube so that supplemental food and nutrition can be delivered through the tube. Such an intervention does not aim to change one's posited life expectancy nor would one be expected to die without it. In this case the juridical threshold of death without intervention would not be met, and obtaining the gastric feeding tube would remain within the realm of permitted actions. Additionally, even if a specific intervention aims at delaying death (setting aside the theological discussions about human agency and its relationship to God's dominion over the specific moment of one's death), that intervention does not have to be indispensable to keeping them alive. Chemotherapy for chronic lymphocytic leukemia might represent such an intervention where treatment could increase life expectancy but at the same time a patient can live for many years without the treatment. Here too, seeking therapy might not be morally obligated according to the juridical stances noted above, and one would be permitted to forgo such therapeutics.

The classical juridical viewpoints described above also implicate clinical decision-making processes in several other ways. For one, these stances illustrate the importance of bringing together clinicians and religious scholars to advise patients and their surrogate decision-makers on courses of action. It is noteworthy that the schools of law that judged there to be an obligation to seek treatment did so provided that (i) the proposed treatment assuredly removes illness-related harms, or (ii) that forgoing intervention results in death. These conditions necessarily implicate epidemiological and clinical data that would be out of the reach of most Islamic scholars. As such, joint consultations between clinicians and jurists are necessary to provide the nuanced guidance about the intended goals of, and the evidence supporting, proposed therapies and about determining the moral valence attached to treatment that rides on the back of these data.

Aside from bringing to the forefront discussions of clinical efficacy and the expected outcomes without treatment, the preceding juridical discussions also subtly suggest that the end-goal of patient-doctor-religious scholar conversations should be the removal of illness-related harms. Some of the preceding Islamic authorities put forth that that the main purpose of medical treatment is the removal of harm (*al-ḍarar*), and harms can be categorized in many different ways. Islamic moral theologians define harm as something that results in a detriment (*al-mafsada*) (Ibn Ḥajar al-Haytamī 2008, p. 516), and in turn a detriment refers to anything that undermines the higher objectives of Islamic law: the preservation of religion (*dīn*), life (*nafs*), intellect (*'aql*), lineage (*nasl*), and wealth (*māl*) (al-Būṭī 2000; Auda 2008). While it is beyond the scope of this paper to discuss in detail, one could argue that the entire healthcare system and all medical practices seek to preserve or restore one or more of these five goods which constitute an Islamic conception of total human well-being. For example, reproductive health procedures and practices aim to protect the prospect of having a lineage, and mental health treatments can be viewed as attempting to preserve intellect.

At the same time, certain elements of medical practice might involve potential harms to one good while advantaging another. Receiving chemotherapy, for example, aims to protect life but risks reducing fertility and thereby is a detriment to lineage. As briefly noted above, Islamic jurists engage in a complex balancing act when determining the ethico-legal status of actions that promote one good while disturbing another because each of these goods have three subcategories (the essential: *ḍarurī*, the necessary: *ḥājī* and the enhancing: *taḥsinī*) as well as private and public dimensions; therefore, there are multiple hierarchies to balance. Islamic law gives precedence to removing harms over procuring benefits (Sachedina 2006), and Islamic notions of harm extend to the afterlife (Arozullah and Kholwadia 2013). Accordingly, in light of the juridical conditions informing the moral obligation to seek treatment and the overarching focus on removing harms, Islamically-inflected, clinical decision-making would bring together patients, clinicians, and jurists to identify what harms a proposed treatment seeks to remove, to assess whether that harm leads to the preservation of the aforementioned five goods, to discern whether the proposed treatment entails harming in order to remove the disease-related harm, and to ascertain what evidence substantiates clinical efficacy for the proposed treatment removing the disease-related harm. By focusing on removing illness-related harms rather than recovery from illness, non-therapeutic, palliative treatments obtain religious utility because they deflect disease-related harms to human well-being.

The juridical stances above restrict the conditions under which medical care becomes obligatory and thereby, albeit somewhat counter-intuitively, open up the space for morally justifiable non-intervention and for non-religious values to hold sway in decision-making. In other words, by restricting the zone of obligation to act, the door is left open for there to be a variety of possible (and Islamically permissible) courses of action, which include forgoing intervention. It is a commonplace (mis)perception that religious individuals tend to seek greater amounts of clinical interventions because their religious values privilege all states of life and demand the usage of "God-given" technological capacities towards preserving life (Orr and Genesen 1997; Doig et al. 2006). This sort of ethical mandate, however, does not hold in a Sunni Islamic context because the moral obligation to seek medical treatment is narrowly circumscribed. Furthermore, the relatively high threshold for conditions to obligate clinical intervention empowers devout Muslim patients and their surrogate decision-makers to calibrate courses of treatment based on a wide range of personal values beyond the religious. Views on one's worth to society and family, tolerance for disability, fiscal responsibility, and many other values can all influence choices about whether to pursue an intervention after it is determined that the conditions that make Islamic authorities consider intervention to be mandatory are not present.

Finally, the aforementioned ethico-legal stances admit a broader ontology of healing that widens the berth for justifiably forgoing clinical intervention. In their discussions about clinical efficacy, jurists (most prominently Ḥanafī authorities) asserted that theology demands that individuals can have their health restored without medical intervention. This notion is advanced on the basis of an Islamic ontology of healing that considers healing to occur by the leave of God, attaches healing qualities to prayer, and also maintains God's prerogative to heal without human intermediaries (Padela et al. 2012; Arozullah et al. 2015; Bakar 2008). Accordingly, forgoing clinical interventions is ethically justifiable in the face of their being alternate paths to healing.

13.3.1 Islamic Conceptions of Human Dignity and Inviolability and the Ethics of End-of-Life Healthcare

Moving upstream from the classical Sunni ethico-legal assessments regarding seeking medical care, there are important theological concepts that can be used to support moral arguments for forgoing clinical interventions near the end of life: *karāmah* and *ḥurmah. Karāmah* and *ḥurmah,* often translated as dignity and inviolability, respectively, are two closely related concepts sourced in the Qur'an and Sunnah that inform ethical thinking about the body and its care. In what follows we will briefly introduce the concepts and then detail how they might impact decisions about clinical interventions near the end of life.

Karāmah derives from an Arabic root that conveys the meanings of honor and generosity (Wehr et al. 1979). Multiple verses in the Qur'an and traditions from the Prophet Muhammad give shape to this Islamic analogue for human dignity. An

Islamic conception of human dignity appears to reside between what Sulmasy terms "attributive" and "intrinsic" models (Sulmasy 2013, 2016). It is attributive in that dignity is conferred by God and given value because of this choice favor, and it is intrinsic insofar as dignity inheres within humankind by the virtue of belonging to the sort of thing that humans are—a special type of God's creatures. The Qur'an stresses that God bestowed dignity to humankind, stating, "We have honored (*karamna*) the sons of Adam... and conferred on them special favors, above a great part of our creation" (Ali 1999; 17:70), and it goes on to further stress the relationship between dignity and God's favor by declaring that "the most honoured of you in the sight of Allah is the most righteous of you" (Ali 1999; 49:13). Importantly Qur'anic commentators note that human dignity is a reflection of God's grace and that all of humanity share equally in this favor; that it is not gained through meritorious conduct, yet further closeness to God is obtained through human actions (Mattson 2002; Kamali 2002). The special dignified rank of humankind is further observed by the rhetorical strategies utilized by the Qur'an to describe the creation of the progenitor human, Adam. Several verses state that God fashioned the human and perfected his image and form (7:11, 40:64, 64:3, 95:4), and in one verse evokes the metaphor of God creating Adam by His hand (38:75) to denote God's care for humanity and the dignified existence bestowed to humankind. Another set of narrations relate that Adam was created in God's image, thereby signifying another level of honor to the human form. And is it on this basis that the Prophet Muhammad forbade disputants to strike one another in the face (Mattson 2002; al-Munajid; Khaṭīb al-Tibrīzī and Robson 1981; al-Naysaburi 2612e, 2841).

Ḥurmah is closely related to *karāmah*; indeed one may argue that *ḥurmah* emerges from *karāmah,* and is derived from an Arabic root that carries the meaning of sacredness and prohibition (Wehr et al. 1979). The term connotes human sanctity, inviolability, and sacredness as gleaned from its usage throughout the Islamic scriptures. Often the term is used to ground negative rights. For example the Qur'an states "Nor take life—which Allah has made sacred (*ḥarramullah*)—except for just cause" (Ali 1999; 17:33). This inviolability of the human extends to his property and honor, as the Prophet Muhammad declared to his followers during his final sermon at the time of Hajj within the sacred precincts of Mecca: "Verily your blood, your property and your honour are as sacred and inviolable (*ḥarām*) as the sanctity (*ḥurmah*) of this day of yours, in this month of yours and in this town of yours" (al-Nawawī, 1524). Indeed the inviolability of the human body extends beyond death as another Prophetic narration reads, "the dignity (*ḥurmah*) of a deceased person is the same as if he or she were alive" (Kamali 2003). Accordingly, the Prophet rebuked a careless grave digger by noting that breaking the bones of the dead is akin to breaking bones of the living (Qasmi 2008; al-Qazwīnī et al.).

Aside from the rulings mentioned by the Prophet as noted above, the concepts of *karāmah* and *ḥurmah* serve as the foundation of multiple ethico-legal injunctions. For example prominent jurists, notably those representing the Ḥanafī school of law, argue that Muslims are duty-bound to protect the integrals of human life irrespective of an individual's creed (Kamali). Juridical authorities also remark that the sanctity of human life extends to the body parts, such that disregarding the sanctity of one part of the human body is akin to disregarding the sanctity of human life itself (Krawietz 2003). Indeed *karāmah* and *ḥurmah* are reflected into Islamic rulings that prohibit the mutilation of the body and disturbing it post-mortem without extenuating legal causes (Qasmi 2008; Krawietz 2003).

So how might these concepts inform decisions about clinical interventions near the end of life? *Karāmah*, insofar as it attaches honor to the human form, and *ḥurmah*, as it considers the human body sacred, give pause to the disruption of bodily integrity and alteration of its appearance. Many types of clinical interventions, by their nature, intrude upon bodily integrity and high-stakes interventions might require advanced monitoring and supportive care mechanisms that also require instruments be placed on or in patient's bodies. Hence decisions about courses of medical care, particularly near the end of life, must balance the posited benefits attached to clinical procedures against the threats to *karāmah* and *ḥurmah* via the violation of bodily integrity and appearance. The notion of *ḥurmah* as it relates to maintain bodily integrity is so critically important to Islamic law that some jurists justify their prohibition of organ transplantation on the basis of it. They hold that the removal of an organ from a body compromises *ḥurmah* and therefore cannot be justified because the surgery brings no benefit to the donor and thus is akin to mutilation. Furthermore, these jurists consider that *karāmah* makes for the "non-usability of human organs [in another body]" (Qasmi 2008). The notion of human dignity was also advanced by Islamic jurists to prohibit organ trade, as it renders the human body as a mere commodity and undergirds Islamic bioethical rulings about the necessity of informed consent in medical treatment and research as dignity demands freedom of choice (Alahmad 2016). Accordingly, healthcare delivery at the end of life that accounts for an Islamic conception of *ḥurmah* would adopt a cautious approach to clinical intervention for "the protection of the bodily integrity and the respectful treatment (*takrim*) of the human body do[es] not merely serve its material quality but acknowledge the superior status of the human being" (Krawietz 2003, p. 197). Harkening back to our discussion about the limited conditions under which clinical intervention is morally obligated by Islam, the theological constructs of *karāmah* and *ḥurmah* provide further support for ethically justifiable non-intervention where interventions have marginal utility and cause great disruption to the body.

The concept of *karāmah* can inform an Islamic ethics of care for the dying in another way; the notion underscores a need to preserve the God-human relationship while delivering healthcare. Since human dignity is conferred by God and nurtured by God-consciousness, healthcare providers should facilitate worship activities that are central to a Muslim's relationship with God and should provide spiritual support through referrals to Muslim chaplains, imams, and other religious leaders as necessary. Bearing in mind that the patient is a dignified creature who has a relationship with God might help healthcare providers to see past the disease that disrupts human physiology and tend to the human spirit that is a receptacle for *karāmah*. For Muslims, dignity is experienced by recognizing one's dependence upon God and acknowledging that illness, health, and cure all come from Him, hence illness does not lessen a patient's dignity but rather may enhance one's experience of it (Mattson

2002). Furthermore, since consciousness is considered a condition for worship as well as for spiritual practices, clinical interventions might be aimed at preserving the cognitive and intellectual faculties insofar as possible such that these religious activities can be maintained. Thus attending to *karāmah* while delivering healthcare at the end-of-life entails attending to the spiritual aspects of the patient's well-being.

In summary the construct of *karāmah* brings the spiritual side of the patient into relief, suggesting that spiritual well-being should be one of the aims of healthcare delivery, while both *karāmah* and *ḥurmah* suggest that the potential for disrupting the sanctity and inviolability of the body should be considered while assessing the benefits and harms of clinical interventions. These concepts are particularly important to consider (and revisit) as patients may end up with ever-diminishing capacities for religious activities and may require ever-increasing amounts of diagnostic and monitoring instruments and life-sustaining measures that are placed on or in their bodies as end-of-life healthcare proceeds.

13.3.2 Searching for a Holistic Islamic Ethics for Healthcare at the End-of-Life

As doctoring has come to involve an increasing use of technology and greater amounts of clinical intervention, many have voiced concern over the erosion of the humanism in contemporary medical practice. The decline of humanistic practice is particularly noticeable in end-of-life care and is contributed to by a variety of social practices and structures within contemporary healthcare (Bishop 2011; Dugdale 2010, 2015). In particular, healthcare has become compartmentalized, as general and specialist physicians, social workers, and chaplains all minister to different aspects of the patient, and all too often take the parts for the whole and overlook the linkages between the patient's psychosocial circumstances, spiritual outlooks, and healthcare choices. Additionally, public discourse on healthcare often portrays physicians as "men against death" and highlights the technological marvels and powers of their practice, consequently making conversations about forgoing clinical interventions seem oddly dissonant to social norms; this can leave patients and providers ill-equipped to discuss and prepare for death (De Kruif 1932; Gordon 2015). To be sure, palliative care as a specialty has secured a room within the house of medicine and seeks to revive conversations about living well while in the process of dying. However, some wonder whether this specialty represents another attempt by medicine to "fix" the "problem" of dying, and whether the noble motivation to help people flourish while dying is obscured by practices such as terminal sedation and "aid in dying" (or assisted suicide) that actually cause death (Bishop 2011; Dugdale 2014; Snyder et al. 2001).

Religious traditions might help modern medicine address these challenges by providing a moral vision for the practice of medicine or at the very least drawing attention to the (humanistic) values that are at stake in the debates about what medicine should be about and what clinicians should offer. In a sense, all of the debates about end-of-life care boil down to two critical questions: (i) What physiological/psychological states constitute a life worth living? (ii) Are healthcare providers morally obliged to use any and all means at their disposal to help patients maintain or to restore a "life worth living"? For example, some individuals may believe that a life of pain or psychological distress is not worth living and thus desire a painless and unconscious decline into death. At the same time, some clinicians might hold that every state of human life is worth living and preserving, or that their professional ethics demands that they do not participate in assisted death. Successfully navigating these potentially conflicting views about patient rights and physician duties has wide-ranging implications for the social ordering of healthcare delivery and the public's expectations of the healthcare system. As we deliberate over how best to attend to care of the dying, Islamic bioethical perspectives are important to develop as they may offer practical guidance to a large portion of humanity—the more than 1.5 billion Muslim individuals and multiple Muslim-majority nations around the globe (Pew Research Center 2009).

As of yet, however, the nascent academic field of Islamic bioethics does not provide answers to the two aforementioned questions in a systematic or holistic way. To begin with, what is a life worth living according to Islam? As discussed above, human life is considered sacred and inviolable by Islam, and the human being is honored and dignified. Given the ability of modern medicine to preserve physiological states of varying degrees with or without affective dimensions of the human, conversations about how these Islamic teachings and theological constructs are reflected in quality-of-life metrics and goals of care near the end of life are still in an emerging state (Padela and Mohiuddin 2015a, b). While there are multiple theological and ethico-legal resources to help identify the states of life that medical therapeutics should aim to preserve, various Islamic bioethical authorities appear to have disconnected views about these matters. Islamic jurists differ on whether abortion is permissible and what conditions justify abortion, and they also disagree about the physiological states that permit withdrawal of life support (Moazam 2005; Brockopp 2008; Sachedina 2017; Qasmi 2008; Krawietz 2003; Padela et al. 2013; Padela and Basser 2012; Yacoub 2001; Atighetchi 2007; Ebrahim 1998, 2008). While ethico-legal pluralism is a valued trait of the Islamic ethico-legal tradition, some of these verdicts are not only contradictory but also overlook considerations about medical practice that render them ineffectual (Padela et al. 2011). One could argue that judgments about when a certain type of life need not be brought into this world and when a certain type of life can be allowed to expire are two sides of the same coin as they attend to a moral vision for what constitutes a life worth living (or a life worth preserving medically). As such, a robust consensus-based theological conception of a life worth preserving in the context of modern medicine (or several such conceptions) would provide a tangible end goal for clinical interventions at the borders of life and death. Additionally, such constructs would provide jurists with an ideal that can be looked to as they mine the sources of Islamic ethics and law to pen rulings that assess the permissibility of intervention and non-intervention at the boundaries of life.

Some scholars suggest that instead of looking to theology, theorization at the level of Islamic ethico-legal genre of the higher objectives of Islamic law, *maqāsid al-shariah*, might provide the end goals needed to delineate ethical healthcare practices. For example, one could argue that the since preservation of life is one of the undisputed higher objectives of Islamic law, and many jurist-theologians have described its essential (*ḍarurī*), necessary (*ḥājī*), and enhancing (*taḥsinī*) aspects, these classifications can be used to develop a bioethical framework for end-of-life healthcare (Saifuddeen et al. 2014). We agree that this strategy has merit, yet feel that theology provides greater space for conceptualization than the *maqāsid* genre does. While a locus of much scholarly attention, scholars continue to debate whether the *maqāsid* are open to revision, whether new *maqāsid* are needed for the modern area, and how reasoning should proceed from *maqāsid* to ethico-legal ruling. While theology is not without its controversy, we believe that moving upstream from ethics and law affords freedom to theorize.

Islamic jurists are keen to limit the zones of obligation (and non-obligation) for clinicians with respect to the question of whether healthcare providers are morally obliged to use any and all means at their disposal to help patients maintain or to restore the state of "a life worth living." As noted above, Sunni authorities considered scriptural evidence as they mapped out the moral status of seeking medical care, and they took care not to attach sin to forgoing treatment by determining this action to be generally permitted. Similarly, jurists have been careful in judging that clinicians are not obligated to provide clinical treatments when such treatments are not efficacious or otherwise not expected to yield positive outcomes. For example, multiple jurists and juridical bodies permit the withdrawal of life support when patients are not expected to recover perception, are terminally ill, or are declared brain dead. These rulings absolve clinicians from the obligation to maintain life support in the face of a vague conception of medical futility (Mohiuddin 2012; Albar and Chamsi-Pasha 2015). While Islamic jurists tend to focus on demarcating the line between *ḥalāl* and *ḥarām* (the question of "can I do such and such?"), we believe that theological conceptualization should accompany these ethico-legal deliberations and would attend to the "should I?" question. In other words, a clinician may not be obligated to withdraw life-support from a patient in a minimally conscious state but should life-support be maintained? While Islamic law does categorize some actions both as recommended (*mandūb* or *mustaḥab*) and as discouraged (*makrūḥ*) and thus can involve the question as to whether one ought to do certain actions, in our reading these categorizations rarely appear in extant Islamic bioethical verdicts. On the other hand, a theological conceptualization of the life worth living (or preserving through medical intervention) that is borne in mind while making ethico-legal assessments would assist all parties in making decisions about courses of clinical intervention and the withholding or withdrawal of advanced life-support.

Furthermore, while Islamic ethico-legal discourse runs rife with discussions on the permissibility of actions, other genres of Islamic ethical reflection address how best to act. For example, the corpus of advice (*adab*) literature focuses on moral formation by inculcating the practices of virtues such that an inner disposition towards the virtuous and meritorious actions results (Sartell and Padela 2015). In addition to *adab* literature, the Islamic tradition also upholds spiritual practices that cultivate God-consciousness and thereby motivate not only acting in accord with the Islamic law but also performing in the way most pleasing to God. A holistic Islamic bioethics should attend to these genres of Islamic ethical reflection that focus on individual moral formation because "being good" coincides with "producing good" (Sartell and Padela 2015). It is obvious that for a holistic ethics of care for the dying, the patient's surrogate decision-makers and providers would benefit from knowledge about how best to accompany the dying and the practices that engender such comportments.

While Islamic theological resources, ethico-legal genres, and spiritual practices exist, what is needed, at least with respect to Islamic ethical reflection about medicine and healthcare, is a way for bringing all of these resources together to offer a holistic, Islamic vision for the practices and uses of medicine. Theological markers of a life worth living and constructs of human dignity and inviolability need to be both foregrounded for Islamic bioethical decision-making and utilized on the back end for calibrating rulings about the permissibility of clinical procedures. At the same time, an Islamic ontology of healing that delineates the roles of patients and providers in attracting healing from God can not only serve as a further check on juridical rulings (as can be seen from the ethico-legal discussions about permissibility of forgoing medical treatments above) but also help to inform advice literature and spiritual practices that support the moral formation of patients, clinicians, and other healthcare actors. Indeed, classical manuals and practices that assist with moral formation might require updating so that they can address the spiritual maladies of a world that is swayed by the power of biomedicine yet persists to have a social ordering of medicine that allows for patient-level health inequities and leaves clinicians with profound questions about the moral worth of their profession. Ultimately, addressing the character and spiritual development of the agents of action, in this case the healthcare actors, is integral to fulfilling the aims of Islamic law—for ethico-legal rulings remain theoretical until embodied within human behaviors. A holistic Islamic ethics for healthcare at the end of life requires bringing together physicians, patients, Islamic jurists and theologians, social scientists, and allied health professionals and stakeholders in a shared enterprise. This enterprise would first focus on generating a more accurate and complete understanding of the ethical problem-space within contemporary healthcare and then seek to generate a conceptual lexicon that allows for these experts to engage in cross-talk to appropriately deploy the methods of each other's fields in deriving Islamic bioethical guidelines and manuals. Such efforts are in a nascent phase, and it is our hope that the rudimentary outline of a methodology that starts with theology and then moves to jurisprudence and brings in moral formation as a complement serves to bolster these fledgling efforts.

13.4 Conclusion

In this paper, we have demonstrated how classical Sunni legal stances on the moral status of seeking medical care and the Islamic concepts of human dignity and inviolability allow for ethically justifying the forgoing of clinical intervention, particularly near the end of life. Hence while the dominant opinions within the Sunni schools of law suggest that we *can* intervene, they in general do not support the claim that we *must*. While research suggests that individuals with religiosity utilize greater healthcare resources at the end of life, that some communities may not accept the withdrawal or withholding of end-of-life care treatments on account of their religious values, and that patients receiving spiritual support from religious communities are more likely to receive aggressive clinical treatments at near the end of life, in our view Islamic perspectives might provide Sunni Muslim patients and families with ethical grounds for less aggressive interventions (Inthorn et al. 2015; Shinall et al. 2014; Shinall and Guillamondegui 2015; Balboni et al. 2013). To be sure, the legal rulings about the moral status of seeking medical treatment vary somewhat across the four Sunni schools of law, have areas of ambiguity, and may not fully account for the nature and epistemic bases of modern biomedicine. Consequently, jurists and juridical councils need to reexamine classical formulae and provide revisions that take into account the deliverables of modern biomedicine and the current social realities of healthcare more fully. Alongside this re-examination, we suggest that Islamic scholars develop a holistic, theologically grounded, Islamic bioethics that brings together theology, law, and spiritual formation, so that both the moral status of actions undertaken by patients, clinicians, and other healthcare actors, and the moral formation of the actor himself or herself is attended to.[4]

References

al-Bukhari, MuḤammad ibn Ismāʿīl. 2002. *Ṣaḥīḥ al-Bukhārī*. Damascus: Dār Ibn Kathīr.
al-Būṭī, MuḤammad Saʿīd Ramaḍān. 2000. *Ḍawābiṭ al-maṢlaḤa fī al-sharīʿa al-islāmiyya*. Beirut: Mu'assasat al-Risāla.

[4]Acknowledgements: AIP's time-effort and research support for OQ was supported by a grant from the John Templeton Foundation (#39623) entitled "Scientific Discoveries & Theological Realities—Exploring the Intersection of Islam and the Human Sciences." Additional time-effort funding for AIP and symposium support was provided by the Doha International Center for Interfaith Dialogue. This paper was presented in partial form at Bayan Claremont College at a symposium entitled "Dignity and Healthcare at the End-of-Life: Abrahamic Faiths in a Bioethics Conversation." The authors acknowledge the support of Dr. Muhammed Volkan Stodolsky who researched Ḥanafī sources. We would also like to thank Drs. Ahsan Arozullah, Faisal Qazi, and Katherine Klima, and Shaykh Mohammad Amin Kholwadia and Jihad Hashim-Brown who were key interlocutors and partners in the working group project that motivated some of the work presented herein.

al-Dardīr, AḤmad ibn MuḤammad, and AḤmad ibn MuḤammad Ṣāwī. 1972. *al-SharḤ al-Ṣaghir 'alā Aqrab al-masālik ilā madhhab al-Imām Mālik*. Egypt: Dār al-Ma'ārif.
al-Khadīm, MuḤammad. 2011. *Ishrāq al-qarār (al-kishkār); Naẓm 'īyādat al-marīḍ; Naẓm al-tafakkur; Naẓm ẓiyārat al-qubūr*. Nouakchott: Ma'had al-Taysīr lil-'Ulūm al-Shar'īyah wa-al-'Arabīyah.
al-Mawṣilī, Abū al-Fadl ʿAbd Allāh b. Maḥmūd. n.d.. *al-Ikhtiyār li-taʿlīl al-mukhtār*. Damascus: Dar Qubāʾ.
al-Munajid, Shaykh Muhammad Saalih. n.d.. 20652: Commentary on the hadeeth, "Allaah created Adam in His image." IslamQ&A. https://islamqa.info/en/20652. Accessed 29 Dec 2015.
al-Nawawī, Abu Zakaria Yahya ibn Sharaf. n.d.. *Riyad as-Salihin*. Sunnah.com. http://sunnah.com/riyadussaliheen. Accessed 29 Dec 2015.
al-Naysaburi, Abu al-Husaya Muslim ibn al-Hajjāj ibn Muslim ibn Warat al-Qushayri. n.d.. Sahih Muslim. *Sunnah.com*. https://sunnah.com/muslim. Accessed 9 Sept 2014.
al-Qazwīnī, Abū 'Abdillāh Muḥammad ibn Yazīd ibn Mājah al-Rab'ī, et al. n.d.. *Sunnah.com*. http://sunnah.com/search/?q=breaking+AND+bone. Accessed 29 Dec 2015.
al-Ramlī, MuḤammad ibn AḤmad. 1967. *Nihāyat al-muḤtaj ila sharḤ al-Minhāj fī al-fiqh 'ala madhhab al-Imām al-Shāfi'ī*. Egypt: MuṢṭafa al-Bābī al-Ḥalabī.
al-Shirwānī, 'Abd al-Hamīd, and AḤmad ibn Qāsim al-'Abbādī. 1972. *Hawāshī al-Shirwānī wa-ibn Qāsim al-'Abbādī 'ala TuḤfat al-muḤtāj*. Beirut: Dār Ṣadr.
al-Sijistānī, Abū Dā'ūd Sulaymān ibn al-Ash'ath. 2009. *Sunan Abī Dā'ūd*. Damascus: Dār al-Risāla al-'Alamiyya.
Alahmad, Ghaith. 2016. Human dignity and its application to medical issues (Arabic). http://www.cilecenter.org/ar/articles-essays/الكرامة-الإنسانية-وتطبيقاتها-في-القض/. Accessed 19 June 2016.
Albar, M.A. 2007. Seeking remedy, abstaining from therapy and resuscitation: An Islamic perspective. *Saudi Journal of Kidney Diseases and Transplantation* 18 (4): 629–637.
Albar, M.A., and H. Chamsi-Pasha. 2015. *Contemporary bioethics: Islamic perspective*. Cham: Springer Open.
Ali, Abdullah Yusuf. 1999. *The Quran translation*. New York: Tahrike Tarsile.
'Alwānī, Ṭāhā Jābir Fayyāḍ, and A.A. Shamis. 2010. *Towards a fiqh for minorities: Some basic reflections*. Herndon: The International Institute of Islamic Thought.
Arozullah, A.M., and M.A. Kholwadia. 2013. Wilayah (authority and governance) and its implications for Islamic bioethics: A Sunni Maturidi perspective. *Theoretical Medicine and Bioethics* 34 (2): 95–104. https://doi.org/10.1007/s11017-013-9247-3.
Arozullah, A.M., M.V. Stodolsky, A.I. Padela, and M.A. Kholwadia. 2015. The role of Muslim ontology in defining a schema of causes and means of healing. Paper presented at the 4th Annual Conference on Medicine and Religion, Cambridge, MA.
Atighetchi, Dariusch. 2007. *Islamic bioethics: Problems and perspectives*. New York: Springer.
Attia, Gamal E. 2007. *Towards realization of the higher intents of Islamic law*. Herndon: The International Institute of Islamic Thought.
Auda, Jasser. 2008. Maqasid Al-Shariah: A beginner's guide. In *Occasional papers series*. London.
Bakar, Osman. 2008. An introduction to the philosophy of Islamic medicine. In *Tawhid and science*, 103–130. Selangor: Arah Publications.
Balboni, T.A., M. Balboni, A.C. Enzinger, K. Gallivan, M.E. Paulk, A. Wright, K. Steinhauser, T.J. VanderWeele, and H.G. Prigerson. 2013. Provision of spiritual support to patients with advanced cancer by religious communities and associations with medical care at the end of life. *JAMA Internal Medicine* 173 (12): 1109–1117. https://doi.org/10.1001/jamainternmed.2013.903.
Bedir, A., and S. Aksoy. 2011. Brain death revisited: It is not 'complete death' according to Islamic sources. *Journal of Medical Ethics* 37 (5): 290–294. https://doi.org/10.1136/jme.2010.040238.
Bishop, Jeffrey Paul. 2011. *The anticipatory corpse: Medicine, power, and the care of the dying*. Notre Dame: University of Notre Dame Press.
Brockopp, Jonathan E. 2008. Islam and bioethics: Beyond abortion and euthanasia. *Journal of Religious Ethics* 36: 3–12.

Dāmād Afandī, 'Abd al-Raḥmān. 1910. *Majma' al-anhur sharḥ multaqā al-abḥūr*. Istanbul: Dār al-Ṭibā'ah al-'Āmirah.
De Kruif, Paul. 1932. *Men against death*. New York: Harcourt.
Doi, Abdur Rahman I. 1984. *Sharī'ah: The Islamic law*. London: Ta Ha Publishers.
Doig, C., H. Murray, R. Bellomo, M. Kuiper, R. Costa, E. Azoulay, and D. Crippen. 2006. Ethics roundtable debate: Patients and surrogates want 'everything done'—What does 'everything' mean? *Critical Care* 10 (5): 231. https://doi.org/10.1186/cc5016.
Dugdale, L.S. 2010. The art of dying well. *Hastings Center Report* 40 (6): 22–24.
———. 2014. Therapeutic dying. *Hastings Center Report* 44 (6): 5–6. https://doi.org/10.1002/hast.379.
———. 2015. *Dying in the twenty-first century: Toward a new ethical framework for the art of dying well*, Basic bioethics. Cambridge, MA: The MIT Press.
Ebrahim, A.F.M. 1998. Islamic jurisprudence and the end of human life. *Medicine and Law* 17 (2): 189–196.
———. 2008. *An introduction to Islamic medical jurisprudence*. Durban: The Islamic Medical Association of South Africa.
Fadel, M. 1996. The social logic of taqlīd and the rise of the Mukhataṣar. *Islamic Law and Society* 3 (2): 193–233.
———. 2008. The true, the good, and the reasonable: The theological and ethical roots of public reason in Islamic law. *The Canadian Journal of Law and Jurisprudence* 21 (1): 5–69.
Farah, Samir, and Ashraf Al-Kurdi. 2006. Brain death: Definition, medical, ethical and Islamic jurisprudence implications. In *FIMA year book 2005–2006*, eds. Hossam E. Fadel, Muhammed A. A. Khan, and Aly A. Mishal, 33–48. Jordan Society For Islamic Medical Sciences.
Ghaly, Mohammed. 2010. *Islam and disability: Perspectives in theology and jurisprudence*, Routledge Islamic studies series. New York: Routledge.
Gordon, M. 2015. Rituals in death and dying: Modern medical technologies enter the fray. *Rambam Maimonides Medical Journal* 6 (1): e0007. https://doi.org/10.5041/RMMJ.10182.
Hamdy, S. 2013. Not quite dead: Why Egyptian doctors refuse the diagnosis of death by neurological criteria. *Theoretical Medicine and Bioethics* 34 (2): 147–160. https://doi.org/10.1007/s11017-013-9245-5.
Ibn Ḥajar al-Haytamī, AḤmad ibn MuḤammad. 2008. *al-FatḤ al-mubīn bi-sharḤ al-Arba'īn*. Jeddah: Dār al-Minhāj.
Ibn MufliḤ al-Maqdisī, MuḤammad. 1996. *al-Ādāb al-shar'īyah*. Beirut: Mu'assasat al-Risālah.
Ibn Qudāmah, Muwaffaq al-Dīn 'Abd Allāh ibn AḤmad. 2007. *al-Mughnī*. Riyadh: Dār 'Ālam al-Kutub.
Inthorn, J., S. Schicktanz, N. Rimon-Zarfaty, and A. Raz. 2015. "What the patient wants...": Lay attitudes towards end-of-life decisions in Germany and Israel. *Medicine, Health Care, and Philosophy* 18 (3): 329–340. https://doi.org/10.1007/s11019-014-9606-5.
Islamic Fiqh Academy. 2000. *Resolutions and recommendations of the council of the Islamic Fiqh Academy 1985–2000*. Jeddah: Islamic Development Bank.
Kamali, Mohammad Hashim. n.d. *Human dignity in Islam*. International Institute of Advanced Islamic Studies (IAIS) Malaysia. http://www.iais.org.my/e/index.php/publications-sp-1447159098/articles/item/36-human-dignity-in-islam.html. Accessed 29 Dec 2015.
———. 2002. *The dignity of man: An Islamic perspective*. Cambridge, UK: Islamic Texts Society.
———. 2003. *Principles of Islamic jurisprudence*. 3rd rev. Cambridge, UK: Islamic Texts Society.
Karman, Karen-Lise Johansen. 2011. The role and the work of the European Council for Fatwa and Research. In *Yearbook of Muslims in Europe*, 655–693. Brill.
Khalid, I., W.J. Hamad, T.J. Khalid, M. Kadri, and I. Qushmaq. 2013. End-of-life care in Muslim brain-dead patients: A 10-year experience. *The American Journal of Hospice & Palliative Care* 30 (5): 413–418. https://doi.org/10.1177/1049909112452625.
Khaṭīb al-Tibrīzī, MuḤammad ibn 'Abd Allāh, and James Robson. 1981. *Mishkat al-Masabih*. Lahore: Sh. Muhammad Ashraf.

Krawietz, B. 2003. Brain death and Islamic traditions: Shifting borders of life? In *Islamic ethics of life: Abortion, war, and euthanasia*, ed. Jonathan E. Brockopp, 195–213. Columbia: University of South Carolina Press.

Mattson, Ingrid. 2002. *Dignity and patient care: An Islamic perspective*. Ziyara: Muslim Spiritual Care Services. https://ziyara.org/resources/dignity-and-patient-care-an-islamic-perspective/. Accessed 29 Dec 2015.

Miller, A.C. 2015. Opinions on the legitimacy of brain death among Sunni and Shi'a scholars. *Journal of Religion and Health* 55 (2): 394–402. https://doi.org/10.1007/s10943-015-0157-8.

Miller, A.C., A. Ziad-Miller, and E.M. Elamin. 2014. Brain death and Islam: The interface of religion, culture, history, law, and modern medicine. *Chest* 146 (4): 1092–1101. https://doi.org/10.1378/chest.14-0130.

Moazam, F. 2005. Islamic perspectives on abortion. *Bioethics Links* 1 (2): 3–4.

Mohiuddin, Afshan. 2012. Islamic perspectives on withdrawing or withholding life support: Ethico-legal responses to moral conundrums. Paper presented at the Islamic Bioethics and Shari'ah Law: The role of traditional scholarship in modern times. The 11th annual Critical Islamic Reflections Conference, Yale University.

Moosa, E. 1999. Languages of change in Islamic law: Redefining death in modernity. *Islamic Studies* 38 (3): 305–342.

Niẓām, Shaykh. 2009. *al-Fatāwā al-hindiyyah*. Beirut: Dār al-Fikr.

Orr, R.D., and L.B. Genesen. 1997. Requests for "inappropriate" treatment based on religious beliefs. *Journal of Medical Ethics* 23 (3): 142–147.

Padela, A.I., and T.A. Basser. 2012. Brain death: The challenges of translating medical science into Islamic bioethical discourse. *Medicine and Law* 31 (3): 433–450.

Padela, A., and A. Mohiuddin. 2015a. Ethical obligations and clinical goals in end-of-life care: Deriving a quality-of-life construct based on the Islamic concept of accountability before god (taklif). *The American Journal of Bioethics* 15 (1): 3–13. https://doi.org/10.1080/15265161.2014.974769.

———. 2015b. Islamic goals for clinical treatment at the end of life: The concept of accountability before God (taklif) remains useful: Response to open peer commentaries on "Ethical obligations and clinical goals in end-of-life care: deriving a quality-of-life construct based on the Islamic concept of accountability before God (taklif)". *The American Journal of Bioethics* 15 (1): W1–W8. https://doi.org/10.1080/15265161.2015.983353.

Padela, A.I., H. Shanawani, J. Greenlaw, H. Hamid, M. Aktas, and N. Chin. 2008. The perceived role of Islam in immigrant Muslim medical practice within the USA: An exploratory qualitative study. *Journal of Medical Ethics* 34 (5): 365–369. 34/5/365 [pii]. https://doi.org/10.1136/jme.2007.021345.

Padela, A.I., H. Shanawani, and A. Arozullah. 2011. Medical experts & Islamic scholars deliberating over brain death: Gaps in the applied Islamic bioethics discourse. *Muslim World* 101 (1): 53–72.

Padela, A.I., A. Killawi, J. Forman, S. DeMonner, and M. Heisler. 2012. American Muslim perceptions of healing: Key agents in healing, and their roles. *Qualitative Health Research* 22 (6): 846–858. https://doi.org/10.1177/1049732312438969.

Padela, A.I., A. Arozullah, and E. Moosa. 2013. Brain death in Islamic ethico-legal deliberation: Challenges for applied Islamic bioethics. *Bioethics* 27 (3): 132–139. https://doi.org/10.1111/j.1467-8519.2011.01935.x.

Padela, A.I., S.W. Furber, M.A. Kholwadia, and E. Moosa. 2014. Dire necessity and transformation: Entry-points for modern science in Islamic bioethical assessment of porcine products in vaccines. *Bioethics* 28 (2): 59–66. https://doi.org/10.1111/bioe.12016.

Pew Research Center. 2009. Mapping the global Muslim population. http://www.pewforum.org/2009/10/07/mapping-the-global-muslim-population/. Accessed 29 Dec 2015.

Qasmi, Mujahidul Islam. 2008. *Contemporary medical issues in Islamic jurisprudence*. New Delhi: Islamic Fiqh Academy.

Qazi, F., J.C. Ewell, A. Munawar, U. Asrar, and N. Khan. 2013. The degree of certainty in brain death: Probability in clinical and Islamic legal discourse. *Theoretical Medicine and Bioethics* 34 (2): 117–131. https://doi.org/10.1007/s11017-013-9250-8.

Rady, M.Y., and J.L. Verheijde. 2013. Brain-dead patients are not cadavers: The need to revise the definition of death in Muslim communities. *HEC Forum* 25 (1): 25–45. https://doi.org/10.1007/s10730-012-9196-7.

Sachedina, A. 2006. "No harm, no harassment": Major principles of health care ethics in Islam. In *Handbook of bioethics and religion*, ed. D.E. Guinn, 265–289. New York: Oxford University Press.

———. 2017. Brain death in Islamic jurisprudence. *Ijtihad Network*. http://ijtihadnet.com/brain-death-islamic-jurisprudence-abdulaziz-sachedina/. Accessed 29 Dec 2017.

Saifuddeen, S.M., N.N. Rahman, N.M. Isa, and A. Baharuddin. 2014. Maqasid al-shariah as a complementary framework to conventional bioethics. *Science and Engineering Ethics* 20 (2): 317–327. https://doi.org/10.1007/s11948-013-9457-0.

Sartell, E., and A.I. Padela. 2015. Adab and its significance for an Islamic medical ethics. *Journal of Medical Ethics* 41 (9): 756–761. https://doi.org/10.1136/medethics-2014-102276.

Shinall, M.C., Jr., and O.D. Guillamondegui. 2015. Effect of religion on end-of-life care among trauma patients. *Journal of Religion and Health* 54 (3): 977–983. https://doi.org/10.1007/s10943-014-9869-4.

Shinall, M.C., Jr., J.M. Ehrenfeld, and O.D. Guillamondegui. 2014. Religiously affiliated intensive care unit patients receive more aggressive end-of-life care. *The Journal of Surgical Research* 190 (2): 623–627. https://doi.org/10.1016/j.jss.2014.05.074.

Skovgaard-Petersen, J. 2015. A typology of fatwas. *Die Welt des Islams* 55 (3–4): 278–285.

Snyder, L., D.P. Sulmasy, Ethics and Human Rights Committee, and American College of Physicians-American Society of Internal Medicine. 2001. Physician-assisted suicide. *Annals of Internal Medicine* 135 (3): 209–216.

Sulmasy, D.P. 2013. The varieties of human dignity: A logical and conceptual analysis. *Medicine, Health Care, and Philosophy* 16 (4): 937–944. https://doi.org/10.1007/s11019-012-9400-1.

———. 2016. Death and dignity in Catholic Christian thought. *Medicine, Health Care, and Philosophy* 20 (4): 537–543. https://doi.org/10.1007/s11019-016-9690-9.

Tavakkoli, S.N. 2008. A comparison between brain death and unstable life: Shi'ite perspective. *Journal of Law and Religion* 23: 24.

Wehr, Hans, J. Milton Cowan, and Thomas Leiper Kane Collection. 1979. *A dictionary of modern written Arabic*. 4 vols. Wiesbaden: Harrassowitz.

Yacoub, A. 2001. *The fiqh of medicine*. London: Ta-Ha Publishers Ltd.

Part IV
Comparative Conclusions

Chapter 14
Comparative Conclusions

Allen Zagoren, Lucy Bregman, Mary Gottschalk, and Timothy D Knepper

Abstract This essay is in fact a set of four mini-essays, each of which was delivered at the final event of The Comparison Project's 2015–2017 series by one of its organizers. Allen Zagoren's conclusion offers us the perspective of a practicing surgeon on the medicalization of death, wrestling with the physician's instinct to save life at all costs, while understanding that this instinct often comes into conflict with what is best for a person and a society. Lucy Bregman's conclusion instead focuses attention on the range of relationships between religious traditions and modern Western medicalized views of death, eventually maintaining that it is not biomedical advances per se that are at issue but rather the role of doctor as medical researcher. Mary Gottschalk's "layperson's" conclusion explores answers to three key questions that emerged over the course of the 2015–2017 programming cycle: (1) Does the fact that we have the medical means to cure disease or prolong life mean that we should do it? (2) What are the guidelines for determining what should be done and when? (3) Who should make this decision and how? Timothy D Knepper's comparative conclusion then ends the essay and volume by attempting to explain the striking similarities between the bioethical positions of different religions by drawing on the cognitive scientific approach of Pascal Boyer.

14.1 Allen Zagoren: How We Die: Evaluation, Reflection, and Prescription

Above all other living organisms, humans have the intellect to anticipate death and to understand that it is inevitable. Many cultures and religions have developed rituals and rules to deal with the process of death as well as prayers, incantations, and

A. Zagoren · M. Gottschalk · T. D. Knepper (✉)
Drake University, Des Moines, IA, USA
e-mail: allen.zagoren@drake.edu; tim.knepper@drake.edu

L. Bregman
Religion Department, Temple University, Philadelphia, PA, USA
e-mail: bregman@temple.edu

T. D. Knepper et al. (eds.), *Death and Dying*, Comparative Philosophy of Religion 2, https://doi.org/10.1007/978-3-030-19300-3_14

rituals to explain and soften the impact of death on those left behind. These rituals allow us to comfort the bereaved, without necessarily providing the framework to contemplate mortality. On the one hand, we fight death and attempt to push it as far into the future as we can, try to ignore its inevitability. On the other, we ponder its meaning, constantly asking: What if anything comes next?

Modern humans face the current dilemma with more sophistication and science than ever. Today, we can reconstruct and replace a significant portion of the human body. We can augment the function of vital organs as they fail and allow them to repair themselves. We can support the interaction of organ systems until a replacement for the failed organ is found and inserted. Interventions include salvage medications and technological innovations such as cardiac pacemakers, respiratory support, antibiotics to treat those infections that in the past portended death for those in the later stages of various diseases,[1] and in the extreme attempts at cardiopulmonary resuscitation (CPR). Our technical sophistication, however, often creates more questions than it answers. Newer technologies aimed at pushing the inevitability of death further into the future have done little more than prolong the process of dying.

Our species continually asks for and receives new therapies and medical advances that create new miracles. These miracles challenge us collectively to put them in perspective as they become commonplace practices, creating a slow disconnect in our reality regarding the inevitable end of life. The following scenario is repeated around the world daily: A loved one is struck with a significant and almost fatal disease. The diagnosis and treatment are vigorous, often painful, usually costly. Despite this, in a quest to stave off death, the patient and family persist until the inevitable occurs. Perhaps a few more months, occasionally a few more years. Despite the modern marvel of therapy and treatment, the patient ultimately succumbs, usually in a hospital or some other impersonal facility.

This oft-repeated scenario has altered the trajectory for healthcare providers and theologians alike. It has created new practices that require people to state their desire to partake of these "end of life" therapies such as a DNR and altered CPR, ventilator status, organ donation, and the like. It has thrust the courts into the fray, as the traditional guidance of religious elders has been transformed into a flurry of court decisions and legal briefs in law and public policy. It has also often put medical technology into conflict with longstanding religious rituals and beliefs. As Michael Gordon, Professor of Medicine at University of Toronto, has observed, "The centuries-old traditions of the gathering of loved ones, the chanting of prayers, the ritual blessings are in the process of being replaced by the 'miracle' of current and modern medical technology" (Gordon 2015).

When should we rely on the complex medical interventions that have replaced the personal rites and rituals, the songs, chants, music, and appeals to the guiding spirits? We must continue to improve our diagnosis and treatment, especially of the chronic diseases that cause life's journey to be more painful and less productive. At the same time, healthcare professionals and spiritual leaders must foster dialogue

[1]As Sir William Osler memorably stated, "Pneumonia is the old man's friend."

about our last days. Can we have the courage to place death in a different dynamic—not one of fear, but one of understanding; not merely comprehending its inevitability, but talking about it, planning for it, and making it part of our life's journey?

Emerging science has put us on the precipice of decisions only previously contemplated in literature and science fiction. The concepts of life everlasting and perpetual youth have been in our literature and art for centuries—challenges posed by authors and philosophers, scientists and visionaries. The concept of life everlasting has been afforded in Western society only to deities. When confronted with such a potential in art and literature, mere mortals eventually succumb, as it is not in G-d's plan for mortals to live forever. Or is it?

As medical care and health advance toward the next millennia, the very concept of life and the potential of transformative life is palpably real. In the current world of social media, we now have the ability to post video recordings that will last as long as the "servers" remain, long after biological death occurs. This concept was once only in the realm of science fiction. Remember Superman's fortress of solitude. First appearing in Superman 17, his parent's memories and consciousness were placed in crystals. Upon returning to the Fortress, he could access the crystals and ask for advice. His parents appeared as holograms and could interact with him and advise him.

Although we are poised on the threshold of discoveries that will make our current technology pale, we are also in the middle of a conversation that should ultimately ask new questions: Can we afford this? The last 10 days of life have a powerful and adverse impact on the healthcare cost-balance sheet. Here is one particularly striking statistic: those at the end of life, who constitute a mere 1% of all patients, account for 30% of healthcare costs (Alemayehu and Warner 2004). Economists tell us that we must get control of this. If we do not, it will cause an economic disaster at some point in the next 5–10 years.

The movie "Logan's Run" depicts a society that has transcended natural death. Babies are created in large incubators. The inhabitants will not die naturally. In order to control the available environment and consumables, the citizens are kept in the dark of this fact. Instead they live their lives, and ultimately when they turn 30-something, they are called to a collective death. Although they are overtly aware of this alteration in their life, it is actually an incinerator that kills them. They are under the belief that life is unending, but there is this transformational event in the future that promises to bring them to another level of life that is more secure and perhaps enjoyable. In its essence, it raises the question: what do we do if we find a way for people to "live forever." How do we ration the finite components of the earth if we double or triple the life span? What do we do and how do we manage life without traditional death?

This is of course the world of science fiction. Or is it? The essence of what we see, feel, touch and experience are processed through the higher portion of brain, the inputs to which are our sensory systems. The cerebrum (and only a fairly small portion) is the computer that stores this information. The rest of the brain simply maintains organ systems and circulation, etc. There is real science that says we will be able to take the images of a dream and memories of a human brain and store them on a computer, that at some point we may even have a reconstructed organic computer (brain) designed simply for storage and maintained by an artificial circulation without a body. As a result, it will be possible to collect thoughts and store them. This is not science fiction.[2] There is even a new word in Webster's lexicon for this new world—*thanatechnology*. Moreover, transhumanist philosophers such as John A. Messerly see transhumanism as the next step in our evolutionary path, a way to create ourselves "more powerful than the gods":

> But if the surge of cosmic longing presses forward, then higher forms of being and consciousness will emerge, and the universe will become increasingly self-consciousness.... Humans are not an end, but a beginning. They need not fear imaginary gods, but need instead to have the courage to create minds more powerful than the gods. Let the dark ages not again descend upon us—let our most fantastic longings be realized. Let us have faith in the future. (Messerly n.d.)

This past year, in Palo Alto in the heart of Silicon Valley, hedge-fund manager Joon Yun started doing a back-of-the-envelope calculation. According to U.S. social security data, he says, the probability of a 25-year-old dying before their 26th birthday is 0.1%. If we could keep that risk constant throughout life, instead of it rising due to age-related disease, the average person would, statistically speaking, live 1000 years. As a result, he, in consort with a number of other investors and health professionals, established the "Palo Alto Longevity Prize." The latest attempt to crack the code of life, it will award $1 million to teams of scientists that demonstrate a reversal of the aging process in test animals. About ten teams have already signed up to compete for the prize, including researchers from nearby Stanford University, as well as the Texas Heart Institute in Houston and Washington University in St. Louis. "We spend more than $2 trillion per year on health care and do a pretty good job helping people live longer, but ultimately you still die," says Dr. Joon Yun, a doctor, investor, and the main backer of the prize. "The better plan is to end health care altogether" (Vance 2014).

In his best-selling book *How We Die*, Sherwin Nuland, American surgeon and Yale University ethicist, presents a series of vignettes about the final journey of his patients and even his family members. In his most widely read text, he discusses the perspective of death on many levels—in some cases omnisciently and in others quite personally. One of Nuland's conclusions is that just as "[e]very life is different from any that has gone before it, … so is every death" (Nuland 1995, p. 3).

What occurs at the moment when the processes of life cease? What does the brain feel? What does it see? What does it hear? Do we smell anything? Like living, dying is a process. It can be measured by seconds, days, or years. Like life, death has a beginning, but as far as we can know, is in itself an end. Today, we face death with more sophistication and better science to help us attempt to answer these riddles.

[2] This is already somewhat of a reality though the Life Extension Institute, a group stores whole bodies or heads in cryopreservation cells. The concept is to preserve the biological vessel until science and technology emerge to bring it back and restore its function.

Let me wax philosophically, if you will, and share my own sense of death as a physician, rather than as an ethicist or academic. As a surgeon, especially one dealing with the critically ill, I was educated to deal with disease. Death was—and still is—the enemy. Like most of my generation, I was trained to ignore the economic impact of the tools I used. Disease and dysfunction needed to be conquered; the cost and the time it took was inconsequential—or at the least the very last thing to be considered. In fact, the ethics of the moment would exclude cost and time. In the trauma bay and operating theater, death is not an option. Overcoming it in the short term is the victory.

I have been confronted with this issue frequently—not daily as many of the television soap operas would lead you to believe, but more frequently than is comfortable. I have seen people die in front of me. I have had their bodies laid open and their organs in my hand, their life's blood ceasing as I stood there virtually helpless. I had to learn quickly, as a young resident and later as an attending surgeon, that I was but a tool of my education and the knowledge before me. I was not omniscient.

I have seen people die slowly and people die quickly. I have witnessed the moment when, as a physician, I had to explain to a patient that, although they felt good, they were entering into the last stage of this life. I have had to do so with the frustration that I could never tell them what might come next. I have watched patients' cellular and biochemical systems slowly shut down, often over months (even though from our first breath we are in essence slowly dying). I have also seen the quick disruption of this process due to outside forces such as massive trauma.

Throughout life, human tissue and organ systems interact at a dizzying pace. This interaction at the chemical level is at once incredibly complicated and amazingly simple. The cells require materials to process into energy. This process results in chemical byproducts (waste), which need to be removed and excreted quickly. There are many invaders attempting to stop the process. It has always amazed me that even under the most dire circumstances, when there is an incredible separation of the cellular status quo, stabilization can occur (often through the body's own defenses) and chaos can subside.

Death too is a complex physiological process. We die in most cases very slowly (at least at the cellular and organ level). The molecular biochemical processes are intricate and difficult to disrupt. Death is not easy to achieve, but when it comes, it is, from the perspective of the physiologist and the pathologist, final.

That, to me, is the incredible mystery—not only that life is a wonder, but also that death is not easy. The mystical is that two situations can exhibit similar and almost uncanny equilibrium—similar circumstances, ages of patients, and relative physiological health—yet one patient survives and the other does not. Despite the application of all of our advanced efforts—technology, training and skill—one lives and the other does not. Ethicists call this "moral luck." I have experienced it, but I cannot explain it. There are forces that we cannot understand and perhaps never will.

As a young child, I often worried about the death of my parents. I would think about it in short bursts, without grasping that it was inevitable. I am fortunate in that

I was able to enjoy my family and my parents into adulthood when, as a physician, I had to stem my fear that my parents too were mortal.

When my father became critically ill, I had to turn off my physician's knowledge and attempt to be a son, brother, and father. I watched with frustration over the course of his death; what were my years of education and training were for if I could not save my father? He only had an infection, not the inevitability of a terminal cancer. When he entered into the last part of his earthly journey, as I knew he would, I watched the machines breathing for him, feeding him, his monitor and heartbeat slowing. Those machines were keeping him alive longer than necessary so as to avoid having him die on my parents' anniversary. I held his hand. And there for a moment—there was a squeeze (or at least I thought there was). This time it was personal.

How many times had I been in a patient's room with medical residents, observing their rituals of death and dying? How many times, upon leaving the bedside, had I taken the moment as a teachable one—learning yet again about the physiology that was ending and our "failure" to produce a "miracle." But this was my father, and I suffered for what seemed the indignity of having his body clearly wasting away, his life force imperceptibly disappearing before my eyes. As I watched the machines breathing for him and feeding him, I realized, perhaps for the first time in my life, not as a physician but as a human, that death, while ultimately inevitable, is sometimes a blessing.

In his best seller, *When Bad Things Happen to Good People*, Rabbi Harold Kushner of Temple Israel in Natick Massachusetts wrote:

> All living creatures are fated to die, but [as far as we know] only human beings know it. Animals will instinctively protect themselves against threats to their life and well-being, but only human beings live in the valley of the shadow of death, with the knowledge that they are mortal, even when they no one is attacking them. The knowledge that we are going to dies someday changes our lives in many ways.... Knowing that our time is limited gives value to the things we do. (Kushner 1981, p. 78)

Let us honor that opportunity by developing a culture that can openly talk about the process and, while not preventing it, help make our journey through life a bit more comfortable for us and those who will live on past us.

14.2 Lucy Bregman: Religions and Medicalized Death

This series of lectures and essays on the topic of religion, death, and medicalized dying has covered the intersections of various traditions with the emergence of new medical technologies. Many of these traditions—religious, philosophical, and cultural—have been caught off-guard by issues such as organ transplants and changing definitions of death. But then, so has everyone. Not even Jules Verne and H.G. Wells, who could foresee powered flight and modern weapons, could ever predict that we could remove a heart from one person and insert it, still alive, into another. In my comparative conclusion, I do not want to focus on the specific responses of religions to these technologies, or the dilemmas of today's dying-prolonging medical care. Instead, I will reflect on some other ways religions can contribute positively to our understanding of death and its place in human existence.

But first, I want to uncover one pervasive assumption made by many who have participated in similar debates and discussions over religion and medicine. That is, the assumption of a "war between science and religion," as if these two were static essentials locked in a battle over who would control the future. For those who use this model, it is obvious which *ought* to control the future: religion's role is rear-guard action, soon to be defeated or made to submit to the overwhelming power and authority of science.

In a memorable illustration of this scenario for the future, I sat through a presentation of The World Values Survey, where the speaker lined up nations on a chart: on one end were the most traditional, at the other end was Sweden. (The USA was quite far behind Sweden, by the way.) For this speaker, the flow of the future was invariably and universally a rush toward catching up with Sweden. Some of us listened in disbelief to this portrayal of cultural imperialism, but I want to evoke it because it lingers in the background of many of our minds. This picture, exemplified in the horizontal chart, assumes a universal linear progress as the direction of human history: the more scientific/rational, the less bound by tradition and influenced or impeded by religion. Eventually, even the nations that drag their feet will arrive at the cutting edge of Sweden-ness. No nineteenth-century advocate of Victorian scientific progress could have been more triumphalist than this presentation, given at an international meeting of a professional society in 2006.

Now, if anything has been learned from this volume of essays drawn from The Comparison Project's lecture and dialogue series on death and dying, I hope it is to undo the certainty of the above version of human historical change. The various religious traditions, which often are portrayed as impeding linear progress, have in reality contributed to it, resisted it, and as we have been shown, often opened up vistas of alternative futures—alternative ways to be in the cosmos and to be mortal humans. Medical practices and hopes have been thoroughly entangled with these complex dynamics. In many varieties, the religious understandings have sought to retain or recover their own versions of *hozho*, the Navaho term for a life of harmony with what is ultimate, and of peace and happiness. Rather than seeing themselves as in competition with Western technologized medicine, religions can point us toward qualitatively different questions and issues as humans encounter death. And they challenge the assumption behind that horizontal chart that we all are going in the same direction to wind up at the same place. Human diversity in culture, values, and worldviews suffers when we imagine such a future timeline.

Some religious traditions have played a direct role in the modernization of the West and in the intentional spread of Western institutions elsewhere. They are in some ways old hands at sponsoring developments in medicine and other Western institutions, and at introducing these in places where they were previously unknown. Although this role can be conflated with colonialism, try to imagine the history of healthcare and education in Africa and Asia without the initial efforts by Christian missionaries. Their work provided infrastructures for these efforts long before national governments were formed, and on which those governments continued to rely after independence. This was mirrored by parallel efforts here in the USA; Roman Catholic nursing orders became the default healthcare system in some Western states, and Protestant denominations strove to provide healthcare for urban environments. In Philadelphia, the names of hospitals recall this tradition: Episcopal, Presbyterian, St. Agnes, etc. No one apparently felt that healthcare was best left to God alone, not even the most Calvinist of believers.

Other religious traditions, indigenous outside Europe and North America, almost always included some healing practices or sheltered the sick in monasteries. However, their relation with Western medicine has been different. They have, starting in the nineteenth century, experienced Western institutions, and their role has been to assimilate, indigenize, and occasionally resist. Perhaps of all Western institutions, scientific medicine has been the least problematic, at least when we compare medicine to Western ideas about law, education, and family life. The essay by Herbert Moyo, for example, reveals the Ndebele people accommodating to both Christianity and their indigenous religion, willing to engage with Western hospitals, but within particular limits such as the rejection of blood transfusions, because a stranger's blood would confuse one's ancestors at the time a person is ready to join them at death. Yet this rejection does not generalize into a "war between science and religion," but reveals how changes and retentions of earlier practices do not follow a straight horizontal line into a universal, set future. Moreover, other major introductions of Western models and institutions caused and continue to cause much more controversy. Questions about women, education, and government have led to far more conflicts than those about medicine. An ironic illustration of this is that the first public speech ever given by a woman in the entire history of Japan was made by the Meiji Empress to raise money for the Red Cross.

But there is a place of resistance for the religions that have now tried to encounter modernity's medicalized death with their own understanding of the human good. It is not over specific medical practices, nor basically over the kind of medical ethics issues (brain death, physician-assisted dying) that have captured media attention. Almost all traditions have endorsed and accepted some version of medicine as a healing practice. Instead, it seems to be Western-model doctors as researchers who are problematic. In fact, Western culture itself has had a problem with this from the early days. Although it was acceptable for Christians to boil the bodies of saints in order to retrieve the bones and distribute them as relics, it was considered a desecration for medical scientists to dig up the dead and perform an autopsy. This squeamishness about turning the person into an object for research, scrutiny, and experiment has haunted medicine, often well outside religious contexts. Mary Shelley's Frankenstein shows the shadow side of science: the doctor's fascination with a scientific puzzle creates a monster and turns him into a moral monster. There is something about dealing with the human body as raw material for someone else's curiosity and tinkering that is deeply troubling for those traditions that take embodiment seriously as a true component of personhood.

How does this apply to actual medical care for the terminally ill? In our anthology, the question asked in the Padela essay, "We can, but must we?" relates back to this. There are situations where medical interventions that we *can* make no longer serve the individual patient, but some other aim. Sherwin Nuland's wonderful classic *How We Die* shows this dilemma in his discussion of "Hope and the Cancer Patient." Although all doctors start as healers who want the best for their patients, it is easy—a slippery slope—to become enamored with the puzzle of the disease itself. While the patient hopes for the cure and puts trust in the doctor, the quest for medical knowledge can leave particular patients behind, and the goal becomes not what is best for the patient but the knowledge that will benefit humans in the future. Nuland showed how the "if we can, therefore we must" mentality carries over into the care of the terminally ill, when increasingly futile treatments are pursued due to a kind of collusion between patient and doctor.

This is not just a psychological problem but one created by this conflation of healer/researcher. The stance of the healer is to view this patient as a human person with an inner awareness and intrinsic value and innate dignity. By contrast, the stance of the researcher requires detached curiosity and, in the worst scenario, an erasure of the patient's humanity for the sake of universally verifiable knowledge. It is possibly the latter that religions have mistrusted, especially when "we can, therefore we must" is assumed as a sign of scientific progress. To be on the Sweden side of that chart, we must disregard our backward-looking qualms and go ahead to further research. The problem, once again, is not an eternal battle between religious authority and scientific knowledge, but a more subtle shift into a mode where humans become means rather than ends in themselves. I do not rest this reluctance to endorse "we can therefore we must" on natural law theory (as Gerald Magill outlines it in this volume), nor did Nuland. But it does rest on some basic intuition about the special dignity of persons, threatened when the lure of "the puzzle" overshadows concern for care.

On a more concrete level, religious traditions also have resources to ameliorate the de-humanizing effects of total institutions such as hospitals. Think of Elliot Dorff's discussion of the Jewish injunction to visit the sick; he draws on traditional injunctions, but recognizes how differently they appear when visits are made in the context of medicalized dying. Much of the discussion of "spirituality in health care" focuses upon this role of softening or blunting the impact of hospital care on patients. But I do not believe religious traditions have resources in reserve to resist completely the triumph of medicalized death, nor that it would be a good thing for them to try. The one case of what appears as total resistance are Christopher Chapple's Jains. They are an instance of a tradition that is so austerely uncompromising in its vision of the proper way to live, that it is hard to imagine them beginning to enter the world of modern western medicine. For people forbidden to ride on or in any vehicle, even a hospital gurney would be beyond the limit. Nevertheless, there something we can learn from these people, as well as other examples such as traditional Africans, that can challenge our ideals of human fulfillment and excellence.

What distinguishes this collection of essays is that they all portray ideals of human fulfillment, ideas of a life well lived, that go beyond a rationalist, instrumentalist, or utilitarian ethic. That ethic was, at bottom, what the presentation of the World Values chart endorsed, and it has become a default for many of us. But this excludes possibilities that may in the long run be of greater value; it shuts them off, blocks them from our vision. We may find we need them, just as we are aware we have let them go. Connectedness with the cosmos—or for many of us, our tie with God—grounds our humanity, and it will be there as strength, memory, and intelligence fade. Next comes our tie to each other, so that any ethic or vision of human purpose that exalts me as an isolated atom with freedom defined by how independent I am from others is defective. Third is our tie to the non-human world of living creatures, for which "stewardship" is the usual term for Christians. Fourth is our tie to the future, which in Muslim tradition is labeled "lineage." My death will not be the end of everything; there will be a world and people who will continue. (What makes the threat of nuclear war so horrible is not just damage to the present but threat to this promise of a populated future.)

I do not see religious traditions as sole custodians of these values, but they can be voices to remind all of us that some ultimate goods matter ultimately. Without these, we will be vulnerable to all sorts of self-serving, short-term goals, to foolish decisions because we assume someone else will be there to pick up the pieces, and to an overriding urge to equate ability to do something with an imperative to go ahead with it—the "we can therefore we must" approach. I see these values as important concerns, which it seems perilous and short-sighted to discount or ignore.

14.3 Mary Gottschalk: Death and Dying: A Layperson's Perspective

Unlike the other contributors to this volume, my engagement with The Comparison Project was not as an academic but as a layperson with an intense personal interest in both comparative religion and bioethics.

On the one hand, this work sharpened my focus on three philosophical questions that sit at the intersection of bioethics and religion:

1. Does the fact that we have the medical means to cure disease and/or prolong life mean that we should do it?
2. What are the guidelines for determining what should be done and when?
3. And perhaps the most important question of all—who should make the decision and how should it be made?

On the other hand, the variety of readings associated with The Comparison Project brought into sharper focus the ethical questions that bedevil both religious authorities and ordinary human beings when they have to deal with medical protocols such as the withholding or withdrawal of medical treatment and physician-assisted suicide (PAS). What we learned from these readings is that, while markedly different religious or philosophical traditions often come to similar bioethical conclusions based on quite different reasoning, there are also significant differences of opinion within most faith traditions, differences that are often based on the underlying source of support for the decision (e.g., revelation vs. reason).

It seems safe to say that most faith traditions have valued good health and encouraged medical care, and for many taking care of the sick was a religious duty. In the Abrahamic tradition, for example, the earliest Western institutions dedicated to care of the sick were established by religious orders; over the centuries, many Christian, Judaic, and Islamic scholars argued that providing good medical care was part and parcel of one's duty to God and one's community.

For much of history, however, good medical care meant little more than proper nourishment and good hygiene. Until the middle of the twentieth century, the level of scientific knowledge was such that many medical problems that now seem trivial were an automatic death sentence. For centuries, whether the patient lived or died was out of the hands of the physician who, like his clerical counterparts, often could do little more than pray or meditate and keep the patient comfortable.

In this environment, life or death for Christians, Jews, and Muslims was seen to be largely in the hands of a Biblical God. For Zulu tribes in southern Africa, life and death reflected the power struggle between the positive intercession of ancestors and the malevolent influence of evil spirits; the efforts of tribal healers or western missionaries would be effective only if the ancestors supported the healing effort. For most Buddhists or Hindus, life and death was largely a function of the karmic law of cause and effect; illness was often viewed as the consequence of negative behaviors in this or a prior life. For many Daoists in China, life and death was the imbalance between yin and yang, a consequence of the failure to live a harmonious life.

Things began to change, at least in Western countries, in the nineteenth century with the discovery, among other things, of germ theory and vaccines. The pace of change accelerated in the middle of the twentieth century, with the development of antibiotics and a variety of surgical interventions that could cure or significantly reduce the mortality from illness and injury.

Many of the religious traditions we studied have been slow to adopt some of these medical innovations and have outright rejected others.[3] This is hardly surprising for at least two reasons.

First, medical innovation has forced patients and their religious counselors to make explicit decisions about matters of life and death that, for most of history, had been out of their control. Within the monotheistic traditions, some view a decision to treat—or not to treat—as an intrusion into or interference with God's divine plan, while others see it as the responsible exercise of human free will and the obligation to use resources wisely (Kelly et al. 2013, pp. 40–41; Dorff 2003, p. 411). For Zulus, Navajos, or Daoists, "modern science" was pitted against traditional healing methods.[4]

[3] Several examples are given in below.

[4] In the first (Zulu) case, see Bozongwana (1983) and Matalane (1989); in the second (Navajo) case, see Cox (2002), pp. 162–165.

Second, the "medicalization of death" has forced patients and their religious counselors to re-think such existential topics as what it means to be alive. Is our humanity in observable phenomena like the capacity for reasoning, the ability to make moral judgments, and the ability to maintain human or spiritual relationships? Or does it lie in faith-based criteria such as the possession of a Christian soul, the Navajo "wind," or the Daoist *qi*? And if you believe in a soul or wind or *qi*, is a brain stem that keeps your hair and fingernails growing sufficient to prove that you are still alive, even if you have no capacity for reasoning or moral judgment?

Those questions loom large today, when machines can maintain breathing and circulation indefinitely for someone in a coma or a persistent vegetative state (PVS)—when artificial nutrition and hydration can maintain the physical body for individuals who can no longer feed themselves or even digest food.

Although different traditions have differing beliefs about the sacredness of life, none of the traditions we explored over the past 2 years were willing to assert that an individual with advanced dementia or in PVS is no longer "alive." None of the traditions we studied approved of euthanasia (mercy killing) or saw PAS as an appropriate solution for someone who has little or no mental capacity.

At the same time, few religious traditions argue that the value of life is absolute—that it must be preserved at all costs. Over the last half-century, increased attention has been focused on the question of whether the burden of medical treatment exceeds any possible or justifiable benefit. The Abrahamic traditions, for example, agree that such a point exists.[5] It is often articulated as the point when you are not choosing to kill the patient but simply letting the patient die. In other words, the fact that we can prolong life does not always mean that we should do it.

Thus, the focus of the debate has shifted to the circumstances under which treatment can or should be withheld or withdrawn, to the circumstances under which—in the words of Catholic Church as far back in the seventeenth century—continuing medical care would be considered "morally extraordinary" (USCCB 2009, pp. 30–31) and therefore not required. A few illustrative examples:

- Within the Jewish tradition, the mitzvah to honor the integrity of the human body (*kavod ha-met*) held pride for place for several thousand years. For some years after organ transplantation came on the medical scene, many rabbinic scholars rejected the concept as a violation of this mitzvah. By 1995, however—perhaps as organ transplants became safer and more reliable—Jewish scholars began to place more weight on the mitzvah of compassion for your community and help for those in need (*besed*) than on the mitzvah of bodily integrity. Even so, many Orthodox Jews still refuse to consider organ transplants. (Dorff 2003, pp. 221–231)
- A different perspective on organ transplants comes from Buddhist traditions, which see one's human existence as integrally linked to the law of karmic cause and effect. In this context, removal of an organ for donation can be seen as potentially giving away some part of one's karmic being. At the same time, Keown

[5] See, for example, Dorff (2003), Kelly et al. (2013), and Padela and Qureshi (2016).

> notes that our humanity is the interaction of five dimensions of existence—physical form, feelings, thought, character, and *viññāna* (consciousness, a sort of spiritual DNA)—but that no one dimension is more important to being human than any other. From this perspective, loss of a single organ might seem of no great karmic consequence. A third perspective, the Buddhist emphasis on compassion, would seem to support the notion of organ donation. Our readings from both Keown (2001) and Tsomo (2006) suggest that, as of the first decade of the twenty-first century, there has been no clear resolution on this issue.

As noted above, none of the religious traditions we looked approved of suicide, which is defined by Merriam Webster as "the act or an instance of taking one's own life voluntarily and intentionally." Until the mid-twentieth century, the connotation of the word was generally that of a promising life cut short for all the wrong reasons. For Western religions, it was taking the decision out of God's hands. For Buddhists, suicide was cutting short an opportunity to earn positive karma; moreover, karmic law suggested that any suffering avoided by suicide might simply revisit you in your next re-incarnation.

Over the last 60 years, however, much ink has been spilled trying to decide whether hastening the time of death for a terminal patient in considerable pain should be considered suicide, morally or legally.[6] From a different perspective, is there a moral difference between hastening death by withdrawing possible treatment and hastening death by injecting a fatal dose of a sedative? Do we need another word—both morally and legally—for this type of voluntary death? One approach came from our study of Jainism, which introduced us to the ritualized process of accelerating the time of death—through incremental fasting—for someone known to be terminally ill. This process was not viewed as suicide but as *sallekhana*, the "thinning out" of life (Chapple 2006).

The third question—who should make the decision and on what basis?—is also front and center as religious traditions attempt to negotiate the distance between their own value systems and secular bioethics. In the latter, both medical ethics and the America legal system state that the decision should be made by the patient, as long as he or she is mentally competent. By contrast, religious bioethics looks to a variety of sources for help in making end-of-life decisions.

The initial source, for most "world religions," tends to be a combination of authoritative texts—think Bible, Qur'an, or Pali Canon—and early commentary. The difficulty for contemporary bioethics is that many of these texts reflect cultures that had very different socio-economic and scientific conditions. Can guidelines developed in a world of extended families, where the "elderly" provided child care and support for working parents, be relevant in a world of nuclear families, where

[6] The debate concerns passive or voluntary euthanasia as well as PAS (e.g., death with dignity), not involuntary euthanasia. Much of the literature supporting this view comes from secular sources, including Quill and Sussman from The Hastings Center. Among our readings, the primary defender was Grayling (2013) (ch. 19), but other religious figures are contemplating "exceptions" to the prohibition on PAS (Kelly et al. 2013, p. 218).

the "elderly" often suffer from chronic physical or mental illness and therefore might need to be maintained in expensive institutional settings?

As a generalization, the more traditional or conservative sects within any religious group tend to apply these historical rules and texts to contemporary situations with little or no alteration. By contrast, religious scholars who question the literal application of ancient texts must turn to theological or philosophical reasoning to determine how ancient guidelines should be applied to current medical issues. All three of the Abrahamic religions have a centuries-long tradition of evaluating ancient texts in a contemporary context, although Islam has been relatively late to adapt this process to the life-and-death issues provoked by medical innovation (Padela 2013). Buddhist scholars have also only recently dipped into the waters of bio-ethics, in part because of the later introduction of medical innovation to many Buddhist communities and in part because Buddhism concerns itself more with individual enlightenment than with issues of social ethics. A similar observation could be made about Daoism. As a general matter, these religious traditions believe that scholars and jurists must examine individual circumstances. Of the religions we studied, only the Catholic Church has a central authority figure that attempts to speak for all members in all circumstances; even under that "magisterial" umbrella, however, Catholic philosophers and theologians offer a wide range of opinions.

The situation is even more ambiguous for religious traditions that do not have canonical texts and are often indistinguishable from the (often changing) cultural context. The conflict between modern medicine and traditional methods of healing for Zulu or Navajo, for example, is not an ethical issue, but a cultural one.

But when all is said and done, virtually all of the religious traditions we explored—including the Catholic Church—acknowledge that in the final analysis, the individual has to make a life-or-death decision based on his or her own reasoning, free will, and/or conscience. Some traditions encourage this more than others—Judaism encourages it, while the Catholic Church discourages it. But it is striking that the fundamental rung of both religious and secular bioethics ultimately rests on the individual.

Within the Abrahamic traditions, there has been a long-standing debate over the potential conflict between God's will—which may be very difficult to discern—and human free will. In terms more applicable to a broader spectrum of faith traditions, it is the distinction between human beings as passive actors in a larger universe and human beings as stewards of the universe we inhabit. If you believe that an essential dimension of being human rests on your capacity and responsibility for stewardship of the world's resources, it becomes increasingly problematic to apply medical solutions developed in very different circumstances to managing life and death for a patient with advanced dementia or terminal cancer, particularly where their continuing care will impose significant emotional, financial, and physical burdens on the immediate family, as well as an already overextended health care system. Dorff captures this well in his comment on "the Jewish tradition's long-standing insistence that individual cases must be decided on their own merits; that general rules may not substitute for careful consideration of the particular circumstances of the people involved; and that, more generally, law and morality are, and must remain, intertwined." (Dorff 2003, p. 411).

Some religious traditions devote more ink to this subject than other, but they all know it is there. Stay tuned.

14.4 Timothy D Knepper: Moral Intuitions, Religious Reasoning, and the Medicalization of Death

During the course of the 2015–2017 programming cycle on death and dying, nothing impressed me more than the striking similarities between different religions about death and dying. To be more precise, it was it was not the "theologies of death" or "rituals of dying" that I found strikingly similar; these, in fact, are quite different across religious traditions. Rather, it was the bioethical positions toward which these traditions argued that I found similar: the permission of withholding and withdrawing life-sustaining treatment, at least in certain circumstances; the rejection of physician-assisted suicide and active euthanasia, in most if not all cases; and the general attempt to strike balanced positions with regard the patient's pain and suffering, the family's social and economic welfare, and the treatment's likelihood of success. The ultimate goal of my comparative conclusion is to explain this similarity. This will be my third and final point. First, though, I begin with two preliminary points, each of which is in its own way contentious, though not as contentious as my final point.

My first point involves generalizations that I, as a philosopher of religion, am trained to resist: death and dying are of concern for most, if not all, religious traditions; religious traditions therefore give reasons about death and dying; and these reasons come in all shapes and sizes. A couple of qualifications and some examples are in order. First, it would be a mistake to project "our" own culture's obsessive concern with the *individual's* death onto all cultures at all times. Not all people were, or even are, preeminently concerned with their own death; not all religion did, or even does, seek to give *meaning* to individual lives in view of their death. Second, for most religious traditions, it seems more important to *ritualize* death and dying than to give reasons about death and dying. What is crucial is that the right practices are practiced, not that the right beliefs are believed. Here too, my "culture" looks strange when compared to others, as rituals are arguably of less and less importance to it.

In the way of providing examples, let me say that when I speak about the many shapes and sizes of religious ideas about and practices of death and dying, what is foremost in my mind are the riches of our series' programming and this volume's contents. We learned about:

- the Tibetan Buddhist view that, after death, the "soul" navigates a between-state for 49 days before transmigrating to a new womb;

- Santa Muerte, the Mexican folk saint of death, who is efficacious for her marginalized devotees, whom the Catholic Church and the Mexican Government have failed;
- *Ars Moriendi*, the medieval Latin texts that detail the temptations, questions, admonitions, and prayers that constitute the "art of dying" for Christians;
- *sallekhana*, a fasting unto death that certain practitioners of the Indian religion of Jainism undertake when failing health prevents them from practicing their religion any more;
- how secular humanists innovate new myths and rituals in order to create meaningful rites of passages at the end of life;
- the long history of Roman Catholic bioethics, especially its use of the Principle of Double Effect to adjudicate the morally ambiguous issues related to withholding and withdrawing;
- the equally long history of Jewish bioethics, and how rabbis draw on Jewish tradition and scripture to offer moral guidance for Jews facing end-of-life decisions about a host of ethical issues;
- the transition from the living-living to the living-dead in the African Ndebele people, and how some of their related practices have run afoul of healthcare professionals;
- the fledgling field of Buddhist bioethics, and how principles such as *karma* affect issues such as organ donation;
- the older and more active field of Islamic Bioethics, and how the teachings of Qur'an and rulings of Sharīʿah impact decisions regarding the removal of futile life-support measures;
- the fear of ghost sickness in traditional Navajo culture, and how health professionals negotiated that fear through traditional Navajo myths, especially as it impacted advance directives; and
- how classical Confucian and Daoist philosophers found meaning in the face of finitude and death without recourse to the notion of personal survival of death.

My second point relates to my own efforts, as Director of The Comparison Project, to "think together" our diverse programming on death and dying. What, for example, is the relationship between formal projects in religious bioethics, on the one hand, and the cult of Santa Muerte, the Jain practice of *sallekhana*, and the Ndebele rejection of blood transfusions, on the other hand? I begin to answer this question by noting that, although the medicalization of death has been a challenge to traditional religious views of death and ways of dying, it is just one part of wider and deeper challenges, all of which we might attribute to modernity in general. To name just a few of these other challenges, as we encountered them in our lectures and essays, I call attention to the following phenomena: socio-economic upheaval, the spread of capitalism and commercialism, access to science and technology, the rise of nation-states and their courts of law, the influence of missionary religions, and waves of pluralism and secularism.

I see three basic kinds of responses to these challenges, at least among our lectures and essays. First, there is *resistance*. Perhaps we saw this best in the Jains defending their right to practice a "fast unto death" in the face of its criminalization by an Indian court of law. But we also see (micro) resistances in Buddhists opting to forgo certain end-of-life medical procedures for the sake of dying as peacefully and consciously as possible, in the Ndebele rejection of blood transfusions, and in the "Death Awareness Movement" and its influence on Christians who want to die as naturally as possible, refusing medicalized death. Second, there is an *adaptation* to challenge. I think we see this most of all in the four bioethical programs that we learned about: Buddhist, Jewish, Catholic, and Islamic. Where organ transplantation was once sometimes problematic, now it is usually not; and where physician-assisted suicide was once entirely rejected, now it is sometimes considered. Third, there is an *innovation* in response to challenge. For me, innovation is seen most clearly in the astounding growth of Santa Muerte devotion, the development of new myths and rituals of death and dying by secular humanists, and the deployment of traditional Navajo myths to support advance-directive initiatives among them.

My third and final point involves some complexity, so I will need to lay some groundwork before getting to it. I begin by looking at what I call the bounds or sources of what is reasonable with regard to religious reason-giving about death and dying. (1) There are bounds or sources that are set by the realities of dying, which include (1a) the physio-psychological realities of pain and suffering; (1b) the bio-medical realities concerning the effectiveness of treatment and chance of recovery; and (1c) the socio-economic realities concerning the availability and affordability of treatment. (2) There are also bounds or sources that are informed by one's cultures and traditions, which include: (2a) whether end-of-life decision-making is addressed by one's tradition, religious or otherwise, (2b) how that tradition is conveyed, whether by some external authority or by subjective interpretation; and (2c) how influenced that tradition is by modernity and other religious-cultural traditions. (3) Finally, there are bounds that would seem to be informed by moral intuitions regarding killing another human being.

This—moral intuition—is for me what is potentially most contentious. It is something that I did not expect to find at the beginning of our 2-year journey into death and dying. It is also something that makes the difference-privileging comparativist in me a little uncomfortable. Nevertheless, nothing struck me more than the uncanny similarities between the bioethical programs of very different religions. On the one hand, Buddhist, Christian, Jewish, and Islamic bioethics (at least as represented in our lectures and essays) all allow for the withholding and withdrawing of life-support measures under certain conditions; on the other hand, these bioethical programs all reject physician-assisted suicide and active euthanasia. Why should religious traditions as different as Buddhism, Christianity, Judaism, and Islam—which argue from tradition-specific principles as different as the law of karma, natural law, divine ownership of the body, and the Islamic theological concepts of human dignity (*karāmah*) and inviolability (*hurmah*)—come to similar positions on end-of-life issues such as euthanasia and withdrawing? The hypothesis that I proffer here is that these uncanny similarities are explained by the facts, so to speak—the realities of death and our moral intuitions about killing—not by the contexts of culture and tradition.

It is the latter—our moral intuitions about killing—that requires explanation. It is often thought that religion is the source of morality. I do not find that view plausible. Rather, I tend to agree with Pascal Boyer's theory that it is people's moral intuitions, among other things, that make religion plausible, not vice versa.[7]

Boyer is a both an anthropologist of religion and a cognitive scientist. On the one hand, he is well aware of the diversity of lived religion throughout the world. On the other hand, he believes that the underlying schematic of religious belief and practice can be explained by the specialized inference systems of the mind. These inference systems were not designed for religious thought and action per se; rather, religious thought and action succeed and breed to the degree that they activate pre-existent inference systems, especially those that are of vital importance to us—those that govern intense emotions, shape our interaction with other people, give us moral feelings, and organize our social groups.

With regard to morality, Boyer maintains that experimental studies show that there is an early-developed specific inference system for morality—a specialized moral sense underlying ethical intuitions. These moral intuitions tend to invoke strong emotions in us, especially when violated. These moral intuitions are also generally realistic and absolutistic—behaviors are seen as either intrinsically right or intrinsically wrong.

This is where the gods and spirits come in. Since our moral intuitions suggest to us that behaviors are right or wrong by themselves, not depending on who considers them or from what point of view, and since we are generally ignorant about where these moral intuitions come from and precisely what they say, we gravitate toward concepts of gods and spirits that explain these moral intuitions. Boyer calls these gods and spirits "full access agents"—beings who have access to all the relevant information about moral situations and therefore know the rightness or wrongness of the relevant behavior. We then take the opinions of the gods or spirits as the source of our moral intuitions, which makes our moral intuitions much easier to represent to ourselves and utilize in further reasoning.

What is the upshot of all this for death and dying? In short, it is that religious reasoning about end-of-life issues might only constitute *post hoc* justifications of decisions that are already made by our moral intuitions and the realities of dying. Let me put this in terms of my three bounds of what is reasonable regarding religious reason-giving about death and dying, this time reordered: First, there are our moral intuitions about killing. Second, there are the realities of dying—pain and suffering, the effectiveness of treatment and odds of recovery, and the availability and affordability of treatment. Finally, there are contextual issues of culture and tradition. What I suggest is that there is much less room for religious tradition in end-of-life decision-making than we might think, that religion must bow, so to speak, to our basic intuitions about killing and the realities of the dying. I am not suggesting that religion plays no role whatsoever, nor that it does not play a significant role in some cases. But I am inclined to think that religious *reasons* generally serve merely as *post hoc* justifications for what has already been decided by our

[7] The following explication draws especially from chapters 4 and 5 of Boyer (2001).

moral intuitions and the realities of the situation. This is my explanation for why the bioethical programs of such different religious traditions should produce similar judgments about the bioethical issues of death and dying. It is what our programming on death and dying has taught me about religious reason-giving.

Of course, there are always exceptions and complications. In the way of exceptions that our programming encountered, I highlight Jainism, for which religious reasons would seem to be paramount. Note, however, that these reasons are deployed in support of views that line up fairly well with the conclusions reached by the four formal bioethical programs: given certain realities of dying, it is morally permissible to withhold/withdraw the support that would otherwise keep one alive. Also note that Jaina philosophers argue strenuously that "*sallekhana* is not suicide," thereby making appeal to moral intuitions about killing.[8]

With regard to complications, I suggest that a "warming" to physician-assisted suicide is faintly detectable in some of our bioethical projects, most notably Gerald Magill's. Can we imagine a future in which physician-assisted suicide is argued as morally permissible by religious theologians and philosophers? Perhaps. What would that mean, then, for so-called moral intuitions about killing? Would we regard the terminally ill, who had voluntarily chosen physician-assisted suicide, as less than human? Or would we regard them the way Jains regard those who partake in *sallekhana*—as people of extraordinary virtue, wisdom, and courage, who choose to end life in a deliberate, reflective, and peaceful manner rather than allowing some disease or illness to reduce them to what is less than human? I close by suggesting, first, that most religious traditions are diverse and supple enough to support physician-assisted suicide both scripturally and theologically; and second, that in most cases of physician-assisted suicide our moral intuitions about killing might not apply (or be triggered), given that the act was voluntarily chosen and conscientiously administered. Perhaps, then, our moral intuitions about killing are as flexible as the theologies in which they get articulated—relative to the realities of death and dying, however they are understood and expressed by religious theologies, which are ways of making sense of our moral intuitions (among other things).

References

Alemayehu, Berhanu, and Kenneth E. Warner. 2004. The lifetime distribution of health care costs. *Health Services Research* 39 (3): 627–642.

Boyer, Pascal. 2001. *Religion explained: The evolutionary origins of religious thought*. New York: Basic Books.

Bozongwana, W. 1983. *Ndebele religions and customs*. Gweru: Mambo Press.

Chapple, Christopher K. 2006. Dying and death: Jaina dharma traditions. In *Dying, death, and afterlife in dharma traditions and western religions*, ed. Adarsh Deepak and Rita DasGupta Sherma, 45–56. Hampton: Deepak Heritage Books.

[8] See especially, Tukol (1976). See also Chapple's essay in this collection.

Cox, Gerry R. 2002. North American native care of the dying and the grieving. In *Death and bereavement around the world*, ed. John D. Morgan and Pittu Laungani, vol. 1, 159–182. Amityville: Baywood Publishing Company.
Dorff, Elliot N. 2003. *Matters of life and death: A Jewish approach to modern medical ethics*. Philadelphia: Jewish Publications Society.
Gordon, Michael. 2015. Rituals in death and dying: Modern medical technologies enter the fray. *Rambam Maimonides Medical Journal* 6 (1): e0007. https://doi.org/10.5041/RMMJ.10182.
Grayling, A.C. 2013. *The God argument: The case against religion and for humanism*. New York: Bloomsbury.
Kelly, David F., Gerard Magill, and Henk Ten Have. 2013. *Contemporary Catholic health care ethics*. Washington, DC: Georgetown University Press.
Keown, Damien. 2001. *Buddhism and bioethics*. New York: Palgrave.
Kushner, Harold. 1981. *When bad things happen to good people*. New York: Random House.
Matalane, Lissah J.T. 1989. *The experiences of death and dying of Zulu patients, their families and caregivers*. Diss. University of Natal.
Messerly, John G. n.d. *A glorious future. Reason and meaning: Philosophical reflections on life, death, and the meaning of life*. https://reasonandmeaning.com/2013/12/21/twenty-first-century-technology/. Accessed 16 Aug 2018.
Nuland, Sherwin B. 1995. *How we die: Reflections of life's final chapter*. New York: Vintage.
Padela, Aasim. 2013. Islamic bioethics: Between sacred law, lived experiences, and state authority. *Theoretical Medicine and Bioethics* 34 (2): 65–80.
Padela, Aasim, and Omar Qureshi. 2016. Islamic perspectives on clinical intervention near the end-of-life: We can but must we? *Medicine, Health Care, and Philosophy*. https://doi.org/10.1007/s11019-016-9729-y.
Quill, Timothy and Bernard Sussman. n.d. *Physician assisted death*. The Hastings Center. https://www.thehastingscenter.org/briefingbook/physician-assisted-death/. Accessed 14 Aug 2018.
Tsomo, Karma Lekshe. 2006. *Into the jaws of Yama, lord of death: Buddhism, bioethics, and death*. Albany: State University of New York Press.
Tukol, T.K. 1976. *Sallekhana is not suicide*. Ahmedabad: L.D: Institute of Indology.
United States Conference of Catholic Bishops (USCCB). 2009. *Ethical & religious directives for Catholic health care services*. 5th ed. Washington, DC: USCCB.
Vance, Ashlee. 2014. Silicon Valley investor backs $1 million prize to end death. *Bloomsberg Businessweek*. http://paloaltoprize.com/media-old/. Accessed 14 Aug 2018.